THE BUSINESS OF LEISURE

The BUSINESS of LEISURE

Tourism History *in* Latin America *and the* Caribbean

Edited by
ANDREW GRANT WOOD

University of Nebraska Press
Lincoln

© 2021 by the Board of Regents of the University of Nebraska

Portions of chapter 6 first appeared in *Making Machu Picchu: The Politics of Tourism in Twentieth-Century Peru*, by Mark Rice. Copyright © 2018 by the University of North Carolina Press. Used by permission of the publisher. www.uncpress.org

All rights reserved

Library of Congress Cataloging-in-Publication Data
Names: Wood, Andrew Grant, 1958– editor.
Title: The business of leisure: tourism history in Latin America and the Caribbean / Edited by Andrew Grant Wood.
Description: Lincoln: University of Nebraska Press, [2021]
Includes bibliographical references and index.
Identifiers: LCCN 2020007575
ISBN 9781496213228 (hardback)
ISBN 9781496223401 (paperback)
ISBN 9781496224088 (epub)
ISBN 9781496224095 (mobi)
ISBN 9781496224101 (pdf)
Subjects: LCSH: Tourism—Latin America—History—20th century. | Tourism—Caribbean Area—History—20th century. | Latin America—Economic conditions. | Caribbean Area—Economic conditions.
Classification: LCC G155.L3 B87 2021 | DDC 338.4/7918—dc23
LC record available at https://lccn.loc.gov/2020007575

Set in Scala OT by Laura Buis.

CONTENTS

List of Illustrations . . vii

Acknowledgments . . ix

Introduction: Travel History's Checkered Past as Prelude to Future Catastrophe? . . 1
ANDREW GRANT WOOD

Part 1. Burgeoning International Travel

1. From the Andes to the Alps: Colombian Writers on Travels in Europe . . 23
MERI L. CLARK

2. Railroads and Steamships: Foreign Investment in the Early Development of Peruvian Tourism, 1900–1930 . . 49
FERNANDO ARMAS ASÍN

3. Changing Caribbean Routes: The Rise of International Air Travel . . 67
BLAKE C. SCOTT

4. From the "Romance of Industry" to the "National Soul": Promoting Travel in the Pan American Union . . 85
ANADELIA ROMO

Part 2. Developing National Tourism

5. The Making of an Elite Tourist Enclave: Viña del Mar's Miramar Beach, Chile (1872–1910) .. 119
RODRIGO BOOTH

6. "To Know Peru Is to Admire It": National Tourism Promotion and Populism in Peru, 1930–1948 .. 147
MARK RICE

7. Domestic Tourism in Golden-Age Veracruz, Mexico .. 171
ANDREW GRANT WOOD

8. The Hotel Casino Project That Put Ecuador's Tourism Hopes on Pause .. 195
KENNETH R. KINCAID

Part 3. Politics, Projects, and Postwar Possibilities

9. An Alliance for Tourists: The Transformation of Guatemalan Tourism Development, 1935–1982 .. 223
EVAN WARD

10. "Created by God" (or Columbus?) for Tourism: Building Tourism Fantasy in the Dominican Republic, 1966–1978 .. 245
ELIZABETH MANLEY

Part 4. Postmodern Ironies and Dark Tourism

11. Mina El Edén and Dark Tourism in Zacatecas, Mexico .. 279
ROCIO GOMEZ

12. Netflix *Narcos* and Narco-Tours: Film Tourism Meets Dark Tourism in Medellín, Colombia .. 299
FÉLIX MANUEL BURGOS

List of Contributors .. 317
Index .. 321

ILLUSTRATIONS

1. John Tallis's map of South America, 1850 . . 4
2. José Clemente Orozco, *Turistas y Aztecas*, 1935 . . 15
3. Victorian hikers on the Chamonix glacier in the Savoy Alps, 1867 . . 32
4. Adolphe Braun, *Swiss Alps above Chamonix*, ca. 1858 . . 33
5. Unidentified artist, *"The English have invaded"* . . 36
6. Neptune party, Grace Line steamer, 1923 . . 52
7. Map of Peru Southern Railways routes . . 54
8. Hiram Bingham III, *Machu Picchu in 1912* . . 57
9. Lima, Oroya, and Huancayo Railroad . . 58
10. Pan American World Airways first passenger flight, 1928 . . 68
11. Cover of *New Horizons* magazine featuring *Yankee Clippers Sail Again* . . 72
12. Pan American World Airways brochure, *The Air-Way to Havana* . . 74
13. Camilo Blas, *Peruvian Country Dance* . . 97
14. *February (Refusing to go in the water)*, from *Zig-Zag* magazine, 1908 . . 122
15. Lifeguard Miguel Pérez, pictured in *Sucesos*, 1916 . . 124

16. View of Miramar Beach, from *Sucesos*, 1916 . . 130
17. Swimmers at Miramar Beach, from *Sucesos*, 1916 . . 131
18. Games at the Valparaíso Sporting Club . . 137
19. Arriving at the beach in Viña del Mar, from *Zig-Zag* magazine, 1905 . . 139
20. Hotel Company of Peru advertisement for government-built hotels, 1941 . . 148
21. Advertisement for Huancayo, Tingo María, and Huánuco hotels, 1944 . . 156
22. Advertisement for Hotel Cusco, 1944 . . 158
23. Advertisement for Hotel Puno, 1944 . . 159
24. Villa del Mar, ca. 1925 . . 173
25. Hotel Diligencias and Veracruz city center, ca. 1950 . . 176
26. Advertisement for Mexico City–Veracruz bus schedules and ticket prices, ca. 1950 . . 181
27. Ignacio Zaragosa, *Hotel Villa del Mar Swimming Pool*, 1950 . . 184
28. Female champion, Veracruz Deep Sea Rodeo, 1948 . . 185
29. Map of Ecuador . . 198
30. Otavalo Plaza . . 201
31. Lake San Pablo . . 214
32. Inti Raymi dancers, Otavalo . . 215
33. Ecuadorian woman . . 217
34. Tikal airfield, 1971 . . 232
35. Tikal temple 1 . . 235
36. Postcard showing La Ceiba de Colón . . 256
37. C. B. Waite, *Historic Downtown Zacatecas*, ca. 1900 . . 282
38. Tourists at silver mine entrance . . 284
39. Pablo Escobar mug shot . . 301
40. Narcotourism graph, 2014–17 . . 304
41. Narco-Tour packages . . 305
42. Colombia narco hoodies . . 306
43. Tourists and guide in Medellín . . 307
44. Medellín neighborhood . . 310

ACKNOWLEDGMENTS

It may have first been Confucius who quite wisely observed, "Wherever you go, there you are." So saying, our informal tourism history collective would not have gotten very far without the support of Bridget Barry, Elizabeth Zaleski, copyeditor Maureen Bemko, and the rest of the University of Nebraska Press team. Thank you and safe travels to you all!

THE BUSINESS OF LEISURE

Introduction

Travel History's Checkered Past as Prelude to Future Catastrophe?

ANDREW GRANT WOOD

As I see it, it probably really is good for the soul to be a tourist, even if it's only once in a while. Not good for the soul in a refreshing or enlivening way, though, but rather in a grim, steely-eyed, let's-look-honestly-at-the-facts-and-find-some-way-to-deal-with-them way.

—DAVID FOSTER WALLACE

The travel industry has typically ranked as one of the leading economic sectors in the global economy. International airlines, cruise ships, hotel chains, beach resorts, and all sorts of package tours are big business and growing at an annual world average rate of 3.3 percent. Contributing to the fast-paced and expansive profile of mass tourism, travelers engage a vast array of communication, commercial transport, and hospitality infrastructures. Tourism spending generates significant revenue across a wide cross section of local, regional, and national economies. In 2017 the World Travel and Tourism Council (WTTC) reported that data gathered from 25 regions and 185 countries estimated that industry revenue amounted to approximately $7.6 trillion (U.S. dollars) for 2016—an estimated 10.2 percent of total global GWP (gross world product). In general employment terms tourism was responsible for approximately 292 million jobs: nearly 1 in every 10 employed worldwide.[1]

Yet, with the expansion of the tourism industry also come a number of disastrous consequences. Rapid growth has reached crisis proportions in any number of popular destinations. Cities such as Venice, Barcelona, Mexico City, Miami, and New York City are being overrun. Traffic to national parks and heritage sites is over capacity.[2] Places where industry has been allowed free rein are becoming "McDisneyized" environments marketed to a mass tourism clientele.[3] Perhaps not surprisingly in this endeavor, huge profits are being pocketed by a relatively select elite while legions of underpaid and underemployed service employees do the thankless day-to-day work.[4]

Environmental costs run high as well. Travel via fossil-fueled transport adds considerable amounts of CO_2 gases to the earth's atmosphere. Tourism's consumption of energy and water and the waste produced are fast degrading local ecosystems.[5] Given the complex economic, social, and environmental conditions made manifest through the practice of contemporary travel, developing sustainable solutions requires careful consideration of industry practices as well as tourism history in order to understand how present-day concerns have developed over time.

Tourism Histories

Scholars typically trace the origins of modern tourism in the West to the European Grand Tour. Initiated in the early seventeenth century and practiced to the mid-nineteenth, the Grand Tour saw mostly well-to-do northern European men—as well as some intrepid women—embarking on a two- to three-year journey across the continent to selected destinations in France (Paris), the Netherlands, Belgium, Germany (Berlin, Dresden), Austria (Vienna), and Switzerland (Geneva, Lausanne), before crossing the Alps into northern Italy (Turin, Milan, Venice, Florence) and finally ending up in Rome for examination of classical and Renaissance cultures. Some even continued farther south, to Naples or other parts of southern Italy, as well as Greece. Accompanying the tourist was often an elaborate entourage consisting of a chaperone, tutor, and team of servants.[6] Serving as a rite of aristocratic passage, the Grand Tour was thought to provide travelers an opportunity to gain exposure to European society and, more particularly, selected works of fine art and scholarship: painting, sculpture, architecture, literature, music, and history. It was understood that the Grand Tour was important training, for many of these same adventurers subsequently went on to assume positions of power in European society and beyond. A signifying practice for the privileged few, touring had powerfully made the world their oyster.[7]

Yet as the nineteenth century progressed, the development of railroad, steamship, and other transport infrastructure opened the way for an expanding market in leisure travel. With this growth, tourism would be an activity undertaken not solely by European elites but also by a growing number of middle-class sojourners. The establishment in Britain of Thomas Cook's business in 1841, for example, gave rise to the first "package" tours, which would soon extend beyond western and southern Europe to the Middle East, Africa, Asia, and the Americas. From their office and retail space (selling guidebooks, luggage, footwear, and other accessories) at Ludgate Circus in London, Thomas Cook and Son facilitated an impressive number of excursions to France, Switzerland, Italy, and Scandinavia, as well as Egypt, Syria/Palestine, South Africa, and other destinations within the British imperial realm.[8] Cook advertising appealed to a fairly wide audience by promoting "cheap, conducted tours . . . with inclusive rates" to and from a variety of sites.[9]

Similarly, travel in the United States began as an elite undertaking but then expanded considerably over the course of the nineteenth and early twentieth centuries. With the advent of the motor car, a growing army of tourists set out for Niagara Falls, Saratoga Springs, the White Mountains of New Hampshire, Yosemite National Park in California, and a host of other popular destinations.

At the same time, a fledgling tourism industry was also taking shape in Latin America and the Caribbean. For international excursionists, safer passage in tropical zones came as the result of shifting medical paradigms (e.g., germ theory) and the implementation of strategic health initiatives such as the use of quinine (as an effective prophylaxis against malaria) and mosquito abatement.[10]

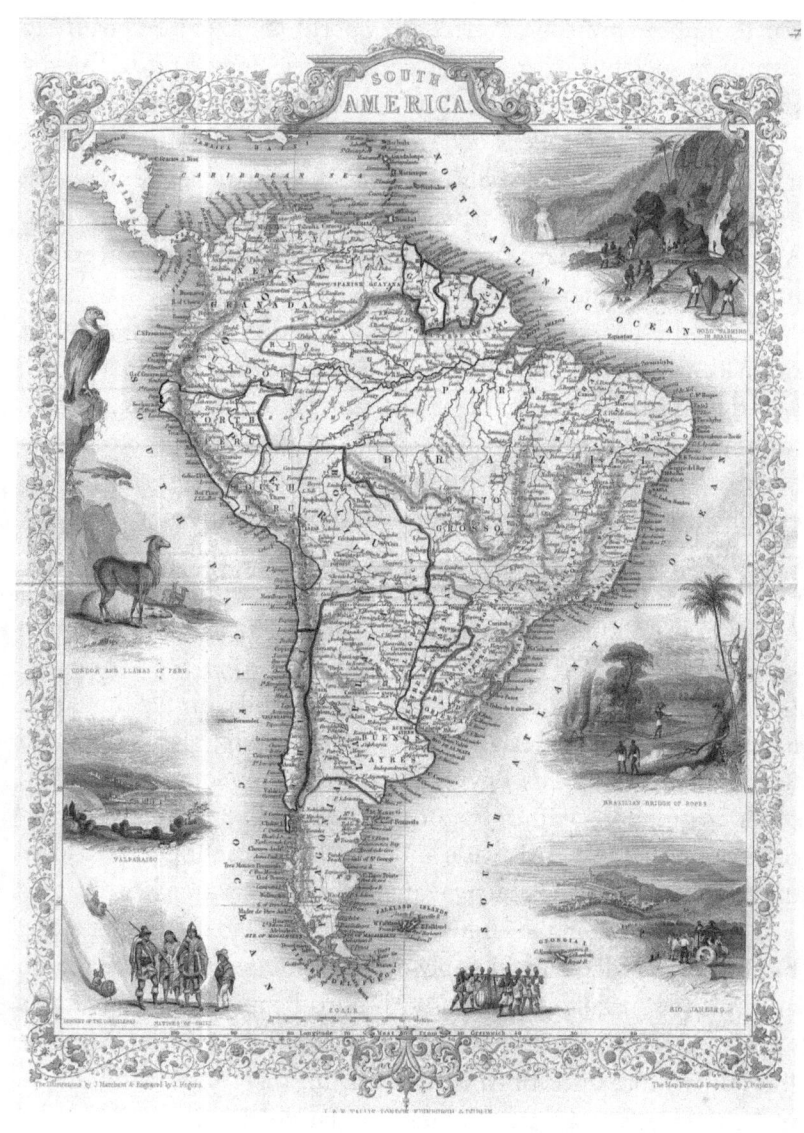

1. John Tallis's map of South America, 1850. Wikimedia Commons.

INTRODUCTION

Prior to the advent of the railroad, "no one traveled for pleasure."[11] Those needing to make passage either hired a litter (two mules harnessed to a padded box outfitted with a leather awning and cotton fabric curtains) or journeyed via horse-drawn coach. Both options were expensive, uncomfortable, time-consuming, and dangerous. Railroads changed this, allowing travelers to bypass many "execrable roads and wretched travel conditions" of the past.[12]

Technology facilitated accelerated movement and commerce and often led to profound social transformation. As the residents of Gabriel García Márquez's fictional Colombian municipality of Macondo were imagined to say, "so many changes took place in such a short time that . . . the old inhabitants had a hard time recognizing their own town."[13] Amazingly, by 1910 rail transport in Latin America and the Caribbean had expanded to include approximately sixty-one thousand miles of track—an amount greater than there was in all of Asia and about three times that of Africa.[14]

The use of steel hulls, refrigeration, and more powerful oil-burning engines provided maritime shipping companies with enhanced opportunities to transport a growing number of passengers. In the 1890s firms such as the New York and Cuba Mail Steamship Company (aka Ward Line) sailed once a week to Havana and the Mexican ports of Tampico, Veracruz, and Progreso. The Ward Line also docked at Nassau in the Bahamas and at Santiago de Cuba every two weeks. The United Fruit Company debuted its "Great White Fleet" in 1899 and subsequently provided Caribbean passage for travelers.

Starting with the Quebec and Gulf Ports Steamship Company, which began offering winter travel from New York to Bermuda and elsewhere in the Caribbean beginning in 1879, the appearance of luxury liners provided an elite alternative to the practice of segregating passenger service by class (i.e., first, second, and "steerage"). These commercial craft featured exclusive, first-class accommodations that included private cabins equipped with running water, along with a variety of recreational areas on board. A U.S. businessman, Henry Plant, further popularized the trend by establishing what is believed to be the first Caribbean passenger cruise in 1891.[15] Subsequently, a host of shipping companies, including Hamburg American, soon redesigned their transatlantic liners for warmer Caribbean waters while Pacific coast operators such as the Grace Line, the Panama Mail, and the Panama Pacific expanded their reach with the opening of the Panama Canal in 1914.[16]

In 1928 Juan Trippe's Pan American World Airways added yet another key technology to the developing tourism industry in the Americas. Initially flying from Key West to Havana, Pan Am soon established regular service between Brownsville, Texas, and Mexico City, as well as another route going from Miami to Havana and then on to the Mexican capital. The company then bought up a number of regional airlines in Central America, purchased landing rights, and acquired seaplane route service between New York, Rio de Janeiro, Buenos Aires, and Santiago, Chile. Amassing a fleet of "flying clippers" (Consolidated Commodore and then Sikorsky s-38 and s-40 flying boats), Pan Am soon dominated hemispheric air travel for the period leading up to World War II.[17]

As tourism took shape across Latin America and the Caribbean, the industry engaged a wide variety of people, places, and social practices. In nearly every nation, industry-related initiatives undertaken by civic and commercial agents provided for a growing assembly of leisure-time opportunities. Gradual development of infrastructure, services, and attractions constructed for the purpose of pleasure travel configured a host of popular routes and cultural practices.

A legion of interconnected local tourism economies rose up in cities and provincial towns, as well as ocean, lake, river, forest, jungle, mountain, and border areas. Industry promoters constructed hotels, resorts, restaurants, and recreational zones while also developing heritage sites and a variety of cultural attractions to draw tourists. Helping to facilitate hospitable passage from one place to the next, service providers packaged tours and hired local guides, translators, drivers, and escorts.

Latin American nations built a number of new, modern highways that would significantly facilitate tourism. Inaugurated in 1936, the Nuevo Laredo–Mexico City Highway (dubbed the Pan American

Highway a year later) filled industry leaders and government officials with great hope that Mexico's burgeoning tourism trade would help lift the nation out of its Depression-era downturn. Toward this end, a host of federally sponsored as well as international commercial and civic organizations, such as the Mexican Automobile Association (AMA), worked diligently to promote tourist infrastructural development geared specifically toward motorists. The association's products and services included maps, guidebooks, automobile insurance, and currency exchange for foreign travelers.[18]

In addition to transport, the availability of hotel accommodations was critical to tourism development in Latin America and the Caribbean. Typically, foreign travelers expected a certain level of service, including private bathrooms and reasonably sanitary facilities. In response to growing demand, a rising number of deluxe lodgings were established. In 1890, for example, the United Fruit Company opened the Titchfield, a luxurious resort for international travelers in Port Antonio, Jamaica.[19] Other sites, such as the Agua Caliente Hotel and Casino in Tijuana, Mexico, and the Hotel Nacional in Havana, Cuba, were built soon thereafter.

Playing host not so much to foreign visitors but to domestic tourists at first, South American beach resorts rose up in coastal areas such as Mar del Plata, Argentina, Piriápolis (just outside Montevideo), Uruguay, and Viña del Mar on Chile's Pacific coast.[20] First established in the mid-nineteenth century, these areas gained widespread popularity in the 1920s and 1930s as road construction provided urban-based motor car enthusiasts increasingly practical and affordable means to undertake leisure travel.[21]

In the meantime Caribbean tourism continued its rapid advance. In Cuba significant infrastructural and real estate development took shape following passage of the 1919 Casino and Tourist Bill, which legitimized gambling and created the Cuban Tourism Commission, whose purpose was to open the island to international tourism. Yet despite auspicious beginnings, Cuba's tourism industry fell somewhat into decline during the Depression and war years, given increased competition with neighboring territories such as Puerto Rico, Jamaica, the Dominican Republic, and Florida.[22]

Increasingly, tourists also made their way to the Andean region. Expanding rail service in Peru made passage to more places possible, while emerging heritage sites such as Machu Picchu and the nearby Sacred Valley near Cusco held great appeal for a wide variety of excursionists.[23] Under a Peruvian state-sponsored program in the 1940s, tourist industry leaders set out to significantly bolster domestic tourism through the development and promotion of a number of new hotels. The endeavor sought to stimulate local and regional economies while at the same time facilitating identification with Peruvian civic culture. Elsewhere a host of national publications such as the Chilean *Vacationer's Guide* (*Guía del Veraneante*) similarly provided prospective Latin American tourists with useful information on not just hotels but parklands, scenic attractions, and more.[24]

Improved access provided by motor vehicles of course played a central role in tourist industry development across the Americas. For their part, Argentinians gained considerable prestige as the Latin American nation with the highest rate of automobile ownership. The Argentine government played no small part in this achievement, as national leaders had developed a network of roads and highways to accommodate the rapidly expanding population of excursionists. Significantly facilitating growth during some of the darkest years of the Great Depression, Argentine highway construction had gone from approximately 2,000 kilometers in 1932 to around 110,000 in 1943.[25]

Since World War II the tourism industry in Latin America and the Caribbean has grown exponentially. Cultural and heritage-oriented travel has flourished, drawing both international and domestic travelers to a wide variety of destinations. "Solidarity" seekers visiting Nicaragua during the 1980s dedicated themselves to a variety of tasks meant to lend support to the fledgling Sandinista revolution before the onset of the Contra War and other complications encouraged them to return home.[26] A decade later political-minded visitors to the highlands of Chiapas, Mexico, sought ways to lend support to the antineoliberal Zapatista insurrection. Subsequently, tourist volunteers who traveled to Haiti in the wake of the 2010 earthquake or to Puerto Rico following the devastation wrought by Hurricane Maria in 2017 brought temporary, "feel good" neocolonial spirit to grief-stricken islanders but not much essential, longer-term relief.[27]

Also growing in popularity are a number of curious, "dark" tourist practices such as travel to various sites where killing and torture occurred during the military dictatorships in the Southern Cone, excursions into low-income neighborhoods (i.e., *favela* tours in Brazil), visits to Caribbean plantations, abandoned silver mines, Colombian drug lord hangouts and other, presumed "exotic" places where unique cultural phenomena can be found (i.e., veneration of Santa Muerte in Mexico).[28] Extreme thrill seekers have traversed the Bolivian "Death Road" between La Paz and Coroico, body-surfed the dangerous coastal currents at Zipolite Beach in Oaxaca, Mexico, ice-climbed at Altamayo, Peru, swum with the sharks in Camagüey, Cuba, and made their way to other shock-inducing, adrenaline rush–producing destinations. Sex tourism is on the rise as well, as power and privilege give way to exploitation and sometimes abuse.[29]

Tourism in Latin America and the Caribbean is distinct from that in other areas of the world because of the region's remarkably diverse array of peoples and cultures. Host populations include literally hundreds of different indigenous groups, with people of African, European, and Asian heritage combining in a kaleidoscopic array. Of course the tourist experience in Latin America and the Caribbean varies widely depending on the status, identity, and intentions of the individual tourist, yet nearly any travel undertaken, no matter how brief, will reveal the fact that American society is deeply patterned by immigration, with the long-standing presence of Native and African American identities profoundly differentiating the hemisphere. Travel, in other words, is different in Latin America and the Caribbean compared to that in Europe, Africa, or Asia because local societies fundamentally comprise truly distinct cultural and demographic elements.

Chapter Outlines

Inspired in part by the pioneering scholarship of John Urry and Dean MacCannell, many insightful studies have focused on economic, social, and environmental aspects of contemporary tourism. Still, histories of the industry and travel experience in Latin America and the Caribbean remain in relatively short supply. Much of the existing literature focuses on international travel during the twentieth century, with particular attention being paid to tourism in Mexico, Cuba, the Caribbean, and to a lesser extent Brazil.

The essays commissioned for this volume draw upon original research focusing on several different Latin American and Caribbean nations, including Brazil, Colombia, Cuba, Chile, the Dominican Republic, Ecuador, Guatemala, Mexico, and Peru. In addition to describing key aspects of industry development in a variety of settings, contributors also consider diverse ways in which histories of travel relate to larger political and cultural questions.

Our opening section, "Burgeoning International Travel," begins with Meri Clark's essay, "From the Andes to the Alps: Colombian Writers on Travels in Europe." In perhaps a surprising and curious manner, Clark's consideration of Colombian tourism deals with renowned writers Soledad Acosta de Samper and her partner, José María Samper, as they journey in Europe during the mid-nineteenth century. As the travelers make their way across various parts of the United Kingdom and Europe, Clark observes that the Sampers' reporting on European landscapes and lifeways not only served to familiarize readers back home with aspects of the European touring experience but also tended to issue a critical, often cautionary, message: Latin Americans should remain vigilant in commanding the fate of their respective nations.

Fernando Armas Asín's "Railroads and Steamships: Foreign Investment in the Early Development of Peruvian Tourism, 1900–1930" explains how a select assortment of turn-of-the-century elites promoted early industry growth in southern and central Peru. As Armas Asín describes, stimulating early infrastructural and hospitality sector development came by way of the private sector rather than through national initiatives. In Peru, foreign shipping and touring companies in conjunction with British-financed Peruvian railroad construction led the way during the first decades of the twentieth century.

Blake Scott's essay, "Changing Caribbean Routes: The Rise of International Air Travel," similarly traces key infrastructural growth by following the advent of Pan American Airways and the growing business of commercial aviation. In writing about this historic shift in transport technology, Scott provocatively contemplates a wider set of social and cultural implications, including the most essential ways in which we experience space and time.

Anadelia Romo's chapter, "From the 'Romance of Industry' to the 'National Soul': Promoting Travel in the Pan American Union," makes insightful use of the *Bulletin of the Pan American Union* to observe changing discursive perspectives on the part of the Pan American Union (PAU) in relation to larger international tourism industry trends and global events. In contrast to previous historical interpretations of the PAU, Romo argues that in promoting travel throughout the hemisphere, the organization served not as an agent of U.S. imperialism but instead as an agent of "assertive national agendas promoted by various Latin American interests."

Part 2, "Developing National Tourism," is dedicated to Latin American domestic travel. Rodrigo Booth's essay, "The Making of an Elite Tourist Enclave: Viña del Mar's Miramar Beach (1872–1910)," critically considers wealthy Chileans' development of an exclusive seaside vacation spot. Identifying a major trend in the tourism industry, the chapter describes the development of Miramar Beach under the auspices of tourism promoter Teodoro von Schroeders and how Chilean elites effectively enclosed the seaside space for their own private pleasure and class identification.

Mark Rice's "'To Know Peru Is to Admire It': National Tourism Promotion and Populism in Peru, 1930–1948" provocatively traces travel marketing to a domestic audience. Particularly important in observing industry development at the national level during the mid-twentieth century was the fact that Peruvian leaders viewed tourism as an effective tool in fashioning a populist appeal. Rice's discussion draws important connections between events in Peru and the larger global community as capitalist, socialist, and fascist states in the 1930s and 1940s took an interest in using tourism to bolster patriotic sentiment.

Further adding to considerations of domestic travel, Andrew Grant Wood's "Domestic Tourism in Golden-Age Veracruz, Mexico," traces the way in which industry boosters sought to attract visitors to the port city during key holiday and vacation times. Wood describes how civic and industry leaders worked diligently to make Veracruz an attractive destination by improving sanitation, constructing new hotels, and developing recreational facilities. Unable to compete with the more glamorous Acapulco on Mexico's Pacific coast, Veracruz nonetheless managed a robust postwar national tourist trade, especially during the Easter holiday, when thousands descended on the city's southern beaches.

Kenneth R. Kincaid's chapter, "The Hotel Casino Project That Put Ecuador's Tourism Hopes on Pause," considers hotly contested efforts to promote tourism in northern Ecuador in the late 1950s. Promoters wanted a hotel-casino to be built on the shores of Lake San Pablo, approximately 134 miles north of the Ecuadoran capital. As Kincaid reveals, planning for the resort grew complicated when the local city council had difficulty securing permission to build, since the site was on indigenous-held land deemed sacred. In January 1959 violent conflict erupted between representatives of the city council and indigenous community members, and the unrest tragically claimed the lives of five indigenes. The incident sparked national debate and effectively derailed further tourist development of the site.

The third part, titled "Politics, Projects, and Postwar Possibilities," opens with Evan Ward's chapter, "An Alliance for Tourists: The Transformation of Guatemalan Tourism Development, 1935–1982." Ward examines the emergence of Guatemala as an international destination up to and in the wake of the U.S.-backed Counterrevolution of 1954. Focusing on the El Petén department, where the archaeological site of Tikal is located, Ward traces the combination of Guatemalan and U.S. government interest in preparing Tikal for mass tourism.

In the Dominican Republic dictator Rafael Trujillo's efforts at developing tourism in the 1950s, including the building of several luxury hotels and hosting of a lavish world's fair, led to only a modest increase in the number of visitors to the country. As Elizabeth Manley details in her chapter, "'Created by God' (or Columbus?) for Tourism: Building Tourism Fantasy in the Dominican Republic, 1966–1978," it was not until Trujillo's successor, Joaquín Balaguer, oversaw critical investment in tourism that the industry really began to take off. Balaguer's twelve-year administration oversaw the consolidation of a profitable tourism agenda as officials established a national program for tourism promoting the republic's natural beauty, colonial heritage, and sex appeal to an international audience. As Manley notes, growth continued through the 1970s and 1980s at a rapid pace and featured the arrival of a number of all-inclusive coastal resorts in the Casa de Campo and Punta Cana areas in the eastern part of the country.

The final part, "Postmodern Ironies and Dark Tourism," considers two examples of contemporary dark tourism. As described by Rocio Gomez in her chapter, "Mina El Edén and Dark Tourism in Zacatecas, Mexico," recent visitors to the Mexican city of Zacatecas are curiously encouraged not only to marvel at the colonial architecture and heritage of the city but also to enjoy a tour of the local El Edén silver mine. The experience has tourists stepping inside a retired mine, dressing up as miners, and wandering through a vast underground labyrinth of rock eventually leading to, of all things, a fashionable underground discotheque. By entering the converted mine, Gomez argues, tourists conjure a form of dark tourism thrill seeking that selectively exploits the mine as a place of entertainment rather than a site of significant human and environmental tragedy.

Félix Manuel Burgos's "Netflix *Narcos* and Narco-Tours: Film Tourism Meets Dark Tourism in Medellín" then critically considers the resurgence of media attention on the international drug trade that has helped created a profitable niche for dark tourism entrepreneurs. As Burgos observes, many of those who make their way to Medellín come with preconceived, often highly romantic ideas. As it turns out, the Netflix series *Narcos* has significantly influenced the tourism business in Medellín by portraying the notorious drug kingpin Pablo Escobar in an alluring, heroic mode. Seeking to cash in on the popularity of Escobar and others like him, various narco-tourism operations have taken an increasing number of drug war dilettantes on visits to infamous prisons, cemeteries, neighborhoods, and private homes. Exactly how and what this practice tells us about the future of tourism in Colombia and elsewhere remains both troubling and uncertain.

Tourism Futures

Most who have previously made the choice to travel (myself included) have tended to journey via jet aircraft, stay in a comfortable hotel, enjoy a nice meal, and move about by fossil fuel–powered vehicles while attending a conference, making a trip to an archive, and/or visiting a handful of recommended scenic sites.

Generally, we have not wanted to stop and contemplate—for too long at least—the ever-widening social inequalities, pervasive corruption, violence, and ecological destruction taking place all over the planet.[30] Travel, as now made painfully clear by the COVID-19 pandemic, has ostensibly become a devil's bargain in which privileged people trade planetary and public health for their own pleasure and personal enrichment.[31] If, as some may argue, there truly are ways to be a tourist without contributing to existing social and environmental problems, we urgently need to know.

We may notice people who appear to have successfully adapted themselves to the reigning global economic order in which elite travel is a leading industry; how could they not? Anticipating the argument that "tourism creates jobs," one cannot deny that the industry has afforded employment for able-bodied cooks, cleaners, clerks, and construction workers while others as well manage to scrape by with their laboring as tour guides, taxi drivers, and service providers. But at the same time, we must acknowledge that the kinds of jobs being created are often menial, insecure, unsafe, and unsatisfying, leaving many hospitality workers underemployed and unable to lift themselves out of poverty.

A while back I walked past a young woman in Cusco, Peru, who was dressed up in traditional costume and wearing a flower in her hair so as to be photographed by tourists for a small fee. Apprecia-

2. José Clemente Orozco, *Turistas y Aztecas*, 1935.
Courtesy Gilcrease Museum, Tulsa OK.

tive of her business acumen, I nodded, smiled, and walked on. Yet as I did, I could not help but think about the many other local people who cannot or choose not to connect with the travel economy. What are they doing while thousands of foreign tourists invade their towns and neighborhoods?

I suspect that many, including those referenced in Jamaica Kincaid's pioneering 1988 book *A Small Place*, are frustrated and resentful about the harsh inequities of the new world order. As Kincaid writes,

> Every native everywhere lives a life of overwhelming and crushing banality and boredom and desperation and depression. . . . Every native would like to find a way out, every native would like a rest, every native would like a tour. But some natives—most natives in the world—cannot go anywhere. They are too poor. . . . They are too poor to live properly in the place where they live, which is the very place you, the tourist, want to go—so when the natives see you, the tourist, they envy you, they envy your ability to leave your own banal-

ity and boredom, they envy your ability to turn their own banality and boredom into a source of pleasure for yourself.[32]

Disparities between people who travel and those who are "toured upon" have of course existed for a long time, yet today we are at a turning point. Even before the COVID-19 pandemic, we could no longer pretend, nor could we afford not to know something of the damage and disservice we do when traveling. Can we now continue to consume disproportionate amounts of the world's resources? Can we persist in turning a blind eye to egregious social and political situations for which we hold some responsibility? Do we go on attending professional meetings and scheduling family vacations in places that require our use of untold volumes of water, fossil fuel, electricity, and other precious resources?

Just as efforts to promote fair trade commodities (e.g., coffee, chocolate, clothing, etc.) have taken shape in recent years, contemporary tourism practices, if they are to go forward, similarly need to be significantly rethought so as to assure human well-being and planetary sustainability. Some may say it is already too late yet, perhaps knowing something about the history of the industry in Latin America and the Caribbean will prove helpful in facilitating desperately needed revolutionary change.[33]

Notes

1. World Travel and Tourism, *Travel and Tourism Economic Impact 2018 World*, https://www.wttc.org/-/media/files/reports/economic-impact-research/regions-2018/world2018.pdf.

2. Charlotte Simmonds et al., "Crisis in Our National Parks: How Tourists Are Loving Nature to Death," *The Guardian*, November 20, 2018, https://www.theguardian.com/environment/2018/nov/20/national-parks-america-overcrowding-crisis-tourism-visitation-solutions.

3. Rojek and Urry, "Transformations of Travel and Theory," 3.

4. Martin Hugo, "Tourism Industry Doesn't Pay a Lot but Hires Recent Grads and the Unemployed, Study Finds," *Los Angeles Times*, June 19, 2017, https://www.latimes.com/business/la-fi-travel-jobs-20170627-story.html.

5. Pattullo, *Last Resorts*.

6. Feifer, *Tourism in History*, 95–134.

7. Sweet, *Cities and the Grand Tour*, 1–64; Chard, *Pleasure and Guilt on the Grand Tour*, 1–39.

INTRODUCTION

8. Thomas Cook, https://www.thomascook.com/.
9. *Cook's Australasian Travellers' Gazette and Tourist Advertiser* 4 (June 1, 1892).
10. Cocks, *Tropical Whites*.
11. Pletcher, "Building of the Mexican Railway," 27.
12. Pletcher, "Building of the Mexican Railway," 26.
13. García Márquez, *One Hundred Years of Solitude*, 246.
14. Topik and Wells, *Global Markets Transformed*, 66.
15. Cocks, *Tropical Whites*, 43–44, 49.
16. Cocks, *Tropical Whites*, 49–59.
17. Asif Siddiqi, "Air Transportation: Pan American; The History of America's 'Chosen Instrument' for Overseas Air Transport," U.S. Centennial of Flight Commission, 2013, https://web.archive.org/web/20090511190314/http://centennialofflight.gov/essay/Commercial_Aviation/Pan_Am/Tran12.htm.
18. Berger, *Development of Mexico's Tourism Industry*, 47–49.
19. Cocks, *Tropical Whites*, 51.
20. Pastoriza, *Las puertas al mar*.
21. Booth, "El Estado Ausente."
22. Schwartz, *Pleasure Island*, 31–33.
23. Rice, *Making Machu Picchu*.
24. Rodrigo Booth, "Turismo y representación del paisaje: La invención del sur de Chile en la mirada de la *Guía del Veraneante* (1932–1962)," *Nuevos Mundo/Mundos Nuevos* (online), posted February 16, 2008, https://journals.openedition.org/nuevomundo/25052.
25. Melina Piglia, "Viaje deportivo, nación y territorio: El Automóvil Club Argentino y los origins del turismo carretera, Argentina, 1924–1938," *Nuevos Mundos/Mundos Nuevos* (online), posted September 16, 2008, https://journals.openedition.org/nuevomundo/40923?lang=en.
26. Babb, *Tourism Encounter*.
27. Katz, *Big Truck That Went By*; Wilentz, *Farewell, Fred Voodoo*; Gabriela Coronado, "Insurgencia y turismo: Reflexiones sobre el impacto del turista politizado en Chiapas," *Pasos: Revista de Turismo y Patrimonio Cultural* (online) 6, no. 1 (2008), www.pasosonline.org; Schuller, *Humanitarian Aftershocks in Haiti*.
28. Williams, "Ghettourism and Voyeurism" (on favela tours); Robb, "Violence and Recreation."
29. Hannum, "Sex Tourism in Latin America."
30. Daly, "Wildlife We See."
31. Jane Dunford, "Things Have to Change: Tourism Businesses Look to a Greener Future," *The Guardian*, May 28, 2020, https://www.theguardian.com/travel/2020/may/28/things-had-to-change-tourism-businesses-look-to-a-greener-future.
32. Kincaid, *Small Place*, 18–19.
33. Barnett, "Influencing Tourism at the Grassroots Level."

Bibliography

Adams, Rachel. "Hipster and *Jipitecas*: Literary Countercultures on Both Sides of the Border." *American Literary History* 16, no. 1 (2004): 57–84.

Aron, Cindy S. *Working at Play: A History of Vacations in the United States.* New York: Oxford University Press, 1999.

Babb, Florence. *The Tourism Encounter: Fashioning Latin American Nations and Histories.* Stanford CA: Stanford University Press, 2010.

Barnett, Tricia. "Influencing Tourism at the Grassroots Level: The Role of NGO Tourism Concern." *Third World Studies* 29, no. 5 (2008): 995–1002.

Berger, Dina. *The Development of Mexico's Tourism Industry: Pyramids by Day, Martinis by Night.* New York: Palgrave Macmillan, 2006.

Berger, Dina, and Andrew Grant Wood, eds. *Holiday in Mexico: Critical Reflections on Tourism and Tourist Encounters.* Durham NC: Duke University Press, 2010.

Booth, Rodrigo. "El Estado Ausente: La paradójica configuración balnearia del Gran Valparaíso (1850–1925)." *EURE Revista Latinoamericana del Estudios Urbanas Regionales* 28, no. 83 (May 2002): 107–23.

Brennan, Denise. *What's Love Got to Do with It? Transnational Desires and Sex Tourism in the Dominican Republic.* Durham NC: Duke University Press 2004.

Brown, Dona. *Inventing New England: Regional Tourism in the Nineteenth Century.* Washington DC: Smithsonian Institution Press, 1995.

Chard, Chloe. *Pleasure and Guilt on the Grand Tour: Travel Writing and Imaginative Geography, 1600–1830.* Manchester, UK: Manchester University Press, 1999.

Cocks, Catherine. *Doing the Town: The Rise of Urban Tourism in the United States, 1850–1915.* Berkeley: University of California Press, 2001.

———. *Tropical Whites: The Rise of the Tourist South in the Americas.* Philadelphia: University of Pennsylvania Press, 2013.

Covert, Lisa Pinley. *San Miguel de Allende: Mexicans, Foreigners, and the Making of a World Heritage Site.* Lincoln: University of Nebraska Press, 2017.

Daly, Natasha. "The Wildlife We See, the Suffering We Don't." *National Geographic*, June 2019, 42–77.

Delpar, Helen. *The Enormous Vogue of Things Mexican: Cultural Relations between the United States and Mexico, 1920–1935.* Tuscaloosa: University of Alabama Press, 1992.

de Santana Pinho, Patricia. *Mapping Diaspora: African American Roots Tourism in Brazil.* Chapel Hill: University of North Carolina Press, 2018.

Feifer, Maxine. *Tourism in History: From Imperial Rome to the Present.* New York: Stein and Day, 1985.

García Márquez, Gabriel. *One Hundred Years of Solitude.* Translated by Gregory Rabassa. New York: HarperPerennial, 1992.

Hannum, Ann Barger. "Sex Tourism in Latin America." *ReVista: Harvard Review of Latin America* (Winter 2002). Available at https://revista.drclas.harvard.edu/book/sex-tourism-latin-america.

Hunter, F. Robert. "The Thomas Cook Archive for the Study of Tourism in North Africa and the Middle East." *Middle East Studies Association Bulletin* 36, no. 2 (2003): 157–64.
Katz, Jonathan M. *The Big Truck That Went By: How the World Came to Save Haiti and Left Behind a Disaster*. New York: St. Martin's Griffin, 2014.
Kincaid, Jamaica. *A Small Place*. 1988. New York: Farrar, Straus and Giroux, 2000.
MacCannell, Dean. *The Ethics of Sightseeing*. Berkeley: University of California Press, 2010.
———. *The Tourist: A New Theory of the Leisure Class*. Rev. ed. Berkeley: University of California Press, 2013.
Merrill, Dennis. *Negotiating Paradise: U.S. Tourism and Empire in Twentieth-Century Latin America*. Chapel Hill: University of North Carolina Press, 2009.
Moss, Jeremiah. *Vanishing New York: How a Great City Lost Its Soul*. New York: Dey Street Books, 2017.
Newsome, David, Susan A. Moore, and Ross Kingston Dowling. *Natural Area Tourism: Ecology, Impacts, Management*. Bristol, UK: Channel View Publications, 2012.
Pastoriza, Elisa, ed. *Las puertas al mar: Consumo, ocio y política en Mar del Plata, Montevideo y Viña del Mar*. Buenos Aires: Biblios Rogamos, 2002.
Pattullo, Polly. *Last Resorts: The Cost of Tourism in the Caribbean*. 2nd ed. London: Latin American Bureau, 2005.
Pletcher, David M. "The Building of the Mexican Railway." *Hispanic American Historical Review* 30, no. 1 (1950): 26–62.
Redclift, Michael. *Wasted: Counting the Costs of Global Consumption*. London: Routledge, 2013.
Rice, Mark. *Making Machu Picchu: The Politics of Tourism in Twentieth-Century Peru*. Chapel Hill: University of North Carolina Press, 2018.
Rivers-Moore, Megan. *Gringo Gulch: Sex, Tourism, and Social Mobility in Costa Rica*. Chicago: University of Chicago Press, 2016.
Robb, Erika M. "Violence and Recreation: Vacationing in the Realm of Dark Tourism." *Anthropology and Humanism* 34, no. 1 (2009): 51–60.
Rojek, Chris, and John Urry. "Transformations of Travel and Theory." In *Touring Cultures: Transformations of Travel and Theory*, edited by Chris Rojek and John Urry, 1–22. London: Routledge, 1992.
Rothman, Hal K. *Devil's Bargains: Tourism in the Twentieth-Century American West*. Lawrence: University Press of Kansas, 1998.
Ruiz, Jason. *Americans in the Treasure House: Travel to Porfirian Mexico and the Cultural Politics of Empire*. Austin: University of Texas Press, 2014.
Saragoza, Alex. "The Selling of Mexico: Tourism and the State, 1929–1952." In *Fragments of a Golden Age: The Politics of Culture in Mexico since 1940*, edited by Gilbert Joseph, Anne Rubenstein, and Eric Zolov, 91–115. Durham NC: Duke University Press, 2001.

Schuller, Mark. *Humanitarian Aftershocks in Haiti*. New Brunswick NJ: Rutgers University Press, 2016.

Schwartz, Rosalie. *Flying Down to Rio: Hollywood, Tourists, and Yankee Clippers*. College Station: Texas A&M University Press, 2004.

———. *Pleasure Island: Tourism and Temptation in Cuba*. Lincoln: University of Nebraska Press, 1999.

Sears, John. *Sacred Places: American Tourist Attractions in the Nineteenth Century*. Oxford: Oxford University Press, 1989.

Strachan, Ian Gregory. *Paradise and Plantation: Tourism and Culture in the Anglophone Caribbean*. Charlottesville: University of Virginia Press, 2002.

Sweet, Rosemary. *Cities and the Grand Tour: The British in Italy, c. 1690–1820*. Cambridge: Cambridge University Press, 2012.

Thompson, Krista A. *An Eye for the Tropics: Tourism, Photography, and Framing the Caribbean Picturesque*. Durham NC: Duke University Press, 2006.

Topik, Stephen C., and Allen Wells. *Global Markets Transformed, 1870–1945*. Cambridge MA: Belknap Press of Harvard University Press, 2012.

Urry, John. *The Tourist Gaze: Leisure and Travel in Contemporary Societies*. Los Angeles: Sage, 1990.

Vanderwood, Paul. *Satan's Playground: Mobsters and Movie Stars at America's Greatest Gaming Resort*. Durham NC: Duke University Press, 2010.

Ward, Evan R. *Packaged Vacations: Tourism Development in the Spanish Caribbean*. Gainesville: University Press of Florida, 2008.

Wilentz, Amy. *Farewell, Fred Voodoo: A Letter from Haiti*. New York: Simon and Schuster, 2013.

Williams, Claire. "Ghettourism and Voyeurism, or Challenging Stereotypes and Raising Consciousness? Literary and Non-literary Forays into the *Favelas* of Rio de Janeiro." *Bulletin of Latin American Research* 27, no. 4 (2008): 483–500.

ONE
Burgeoning International Travel

1

From the Andes to the Alps

Colombian Writers on Travels in Europe

MERI L. CLARK

Renowned Colombian authors Soledad Acosta de Samper (1833–1913) and José María Samper (1828–88) had lengthy and varied careers that influenced generations of Latin American readers. Samper published for more than forty years, while Acosta outlived him and wrote for more than two decades after his death. This chapter focuses on an early period of their travel writing from Europe. As journalists for South American newspapers, the couple was based in Paris and London from 1858 to 1862. They reported current events as well as their views of European life and culture. Acosta in particular held Switzerland in high regard. She supposed that Colombia and Switzerland shared a geographic and even cultural affinity. She and her husband set off on a summer excursion to the Swiss Alps, testing that idea. Over a monthlong journey Acosta and Samper reflected on their national identity, their relationship, and even their notions of selfhood as they evaluated the Alpine milieu. Both wanted to educate their readers about Europeans' characteristics, as well as entertain them with stories of foibles (their own and others). Ultimately, Samper's reports lauded two aspects of the trip: his educated, resilient, and "angelic" wife and certain models of industrial progress. He wished readers to emulate his wife, but he warned them against adopting any developmentalist model without careful thought. Acosta was even more circumspect than her husband was. In particular she condemned the growing influence of the English middle class. Her frank, sometimes funny observations mixed with practical advice for readers, bridging the gulf between aspiration and reality. Forewarned was forearmed, they advised future South American visitors to Europe.

Elites of the burgeoning industrial age like Samper and Acosta saw their political, commercial, religious, and leisure opportunities to travel grow as transportation and communication networks improved. The Samper-Acosta family maintained its principal residences in Paris and London, but husband and wife traveled extensively on the continent and in England. They joined a growing number of well-to-do tourists taking part in what was known as the Grand Tour of Europe. Young upper-class European men, and some women, toured Italy, France, Spain, and Germany to complete their education. Urban Switzerland had long figured in the Grand Tour, but Alpine tourism was still a developing market in the 1850s, when Samper and Acosta visited. Steamships and railways lowered the costs of travel, public education expanded with the growth of public museums and libraries, and multigenerational family trips allowed women to travel abroad as never before.[1] Members of the nineteenth-century European middle classes increasingly could afford such excursions; Latin American visitors were rarer.

Travel writing and elite tourism are themselves understood to be an expression of European cultural domination over other lands and peoples. In this nineteenth-century period travel writing boomed as readers sought information and vicarious pleasure from seasoned adventurers. Most authors were upper-class urban men who assumed a sense of cultural superiority in rendering foreign places and people as exotic and uncivilized.[2] Given this, it is less common to consider nineteenth-century Latin American tourists to be a part of this literature. Latin Americans abroad are typically included in studies of exile, with most focused on twentieth-century Latin American émigrés in Europe.[3] Other influential work has drawn our attention to the political dynamics of travel to Latin America.[4]

What happened when Latin Americans turned the tables to write about Europeans? As columnists for South American newspapers,

Samper and Acosta played roles as advocates, critics, ambassadors, and voyeurs. The couple traveled by choice rather than by force, making them among the very few Latin Americans who had the resources of time, capital, and social networks to do so. They advised Latin Americans to adopt sensible goals for commercial, industrial, and social development. They warned against anglo- and francophilia. They prodded Latin Americans to examine their own histories, ecologies, and societies to identify their own paths toward national development. Their affinity for Switzerland as a comparable mountain society informed their writing. They did not seek to copy European culture but tried, through humor, criticism, and measured praise, to share with their Latin American readers both the benefits and costs of the emerging new industrial order. As Colombians, they sought to find a place alongside their European companions, not below them.

Colombian Writers at Home and Abroad

Nineteenth-century Colombia's social unrest, partisanship, and civil wars shaped the lives of Samper and Acosta in many ways. Soledad Acosta was born into unusual privilege in the Colombian capital, Bogotá, but educated abroad. Her mother was Scottish Canadian, and her father, General Joaquín Acosta, was an important revolutionary army officer who became an ambassador and respected historian. They sent the young Acosta to school in Halifax, Canada (1845–46), and then Paris, France (1846–48), until the 1848 revolution forced her to return to Colombia.

Compared to Acosta's globe-trotting childhood, Samper's youth was more conventional for an elite Colombian man. Born in Honda, in north-central Colombia's *tierra caliente* (hot land), Acosta completed his secondary education in Bogotá at his father's behest. There the fifteen-year-old earned his reputation as a liberal with his first published essay, in *El Día* (Day) in 1843. Samper also earned his law degree (1846) and developed his political and professional network in Bogotá as a journalist critical of conservative and clerical politics. In the late 1840s he worked in several private- and state-sector jobs, including as editor of the state newspaper and in the treasury and foreign affairs departments. In 1852, after less than a year of marriage, Samper's first wife, Elvira, died. His grief propelled him from the capital into months of seclusion in his home region, after which he had recovered enough to establish his own law firm and a tobacco company. When General José María Melo was ousted from the presidency in 1854, Samper's political career began in earnest, as he entered the national congress, representing the Colombian province of Panama.

During this period of national unrest, Samper and Acosta met at a festival in Guaduas, Colombia. She declared it love at first sight, and they wed months later, on May 5, 1855. They had two daughters in quick succession: Bertilda on July 31, 1856, in Bogotá, and Carolina on October 15, 1857, in Guaduas. Samper decided to move the family to Europe for a sabbatical from the "passionate politics that asphyxiated him." He felt he was "leaving the Republic at peace and taking advantage of the time to educate" himself in "very cultured countries" with the aim of "serving the Patria better," he later wrote in a memoir.[5] Samper, Acosta, the two little girls, and Acosta's widowed mother, Carolina Kemble, all sailed to London in 1858. The Peruvian newspaper *El Comercio* (Commerce) soon hired Samper and Acosta as European correspondents. In the family's four years abroad, two more daughters arrived. Acosta gave birth to María Josefa on November 5, 1860, in London, and to Blanca Leonor on May 6, 1862, in Paris.[6]

Samper traveled extensively on the continent by himself; Acosta joined him at times. They both published reports and essays in serial for South American readers. Acosta wrote for the Bogotá-based *El Mosaico* (Mosaic), "a magazine for youth, exclusively dedicated to literature," and for *Biblioteca de Señoritas* (Young ladies' library), the first journal directed to female readers in Colombia.[7] Acosta's often front-page columns, "The Parisian Review" and "Dictionary of Curiosities," were the longest-running in the sixty-seven eight-page issues of *Biblioteca* (1858–59). Her early international experiences, as well as her familiarity with national and international politics, likely informed her reflections on European leisure culture. In 1862 *El Comercio* appointed Samper the journal's executive editor and the family moved to Lima, Peru. A year later they returned to Bogotá, where Samper was elected to the national congress of Colombia. Samper positioned himself as a citizen, traveler, and teacher who examined history, society, and development at home and abroad.

Samper's credibility as a foreign observer depended, he thought, on his intelligent analysis and entertaining narration of his inaugural voyage abroad. His first reports from his European travels appeared in serial in *El Comercio*, the Lima newspaper. After the family's return to South America in 1862, these were published as a two-volume book, *Viajes de un colombiano en Europa* (Travels of a Colombian man in Europe). His introduction to *Viajes* contrasted the simple pleasure trips of an incurious "pseudo-traveler" with knowing "how to travel," and undertaking such travels was, he wrote, "more delicate and difficult than it appears." Samper saw himself as a master of this "complex art of methodical investigation and . . . intelligent caprice." He aimed to teach his readers how to travel as educated tourists. He listed artistic features of the well-prepared traveler ("bright impressionability, poetic imagination, stern judgment, curiosity in observation, and freedom of spirit"), as well as practical ones ("time, money, patience, knowledge of languages").[8] He lamented that the "common masses" had no understanding of "European progress," and he seemed to think they never would. Yet even his elite readers would receive such education with a good dose of entertainment, he thought, so his "rapid, faithful and animated narrative" aimed at dispelling misunderstandings about Europeans.[9]

Samper styled his essays as conversational pieces, mixing history with quasi-anthropological observations about presumed national "types." He detailed the small absurdities and humorous cultural contrasts of his fellow passengers and the cities he explored. *Viajes* began by inviting the reader onto the deck of a steamship sailing the Magdalena River, Colombia's main waterway to the Caribbean Sea.[10] He urged readers to appreciate the "sublime" of their own nature: the non-European, non-elite cultures of Latin America. *Mestizaje*, the mixture of European, indigenous, and African peoples, had made Spanish America different and, he figured, more democratic than Europe.[11] Herein lay the nineteenth-century Colombian modernizer's dilemma: How does one learn from Europe without being diverted from one's true national identity? To this, Samper answered, "The democrat of Colombia needs to nourish his spirit with the old civilization's light and strengthen his republican spirit with the severe lessons of a society deeply ulcerated by oppression and privilege."[12] Latin Americans should think critically about European civilization.

The first volume of *Viajes* kept readers close by Samper's side as he sailed down the Magdalena River to the Caribbean and then crossed the Atlantic. After arriving in England in 1858, Samper set off on a three-month tour of Spain without his family, reporting his trip for *El Comercio*. His writing fit the literary genre of *costumbrismo*, which had emerged in 1830s Spain and observed customs, manners, folklore, and social life. His colorful shipboard vignettes offered spurious stereotypes in a weak attempt at humor, pigeonholing Europeans in ways that would have been familiar to readers. For example, he described a sixty-two-year-old Irishman as "big like a tower, happy like a boy, drinker of first order, as was his obligation to honor his nationality, and a playful joker like all the Irish (except the serious ones), who introduced a delicious disorder on deck."[13] Samper pandered to stereotype again in describing his English roommate, who had stymied his every attempt to communicate: "For many days I longed to strike up a conversation, forgetting that, if I were expansive as a Colombian-Spanish gentleman, my neighbor was of the taciturn and ceremonious race of John Bull. All I could pull from him after five days was *thank you, sir*, spoken dully, for having moved a plate of oranges toward him."[14] Returning to the ship after another absurdly silent encounter with the Englishman at a tourist site in northern Spain, Samper began "writing a few verses for my wife," vowing that such "silence would never enter my life habits."[15] Rather than trying to decipher these foreign characters, cultures, and values, Samper turned his reflections inward. His wife served as his South American audience writ small. To her, he wrote that "we" were right to attend generously to the aloof English, the drunken Irish, and others, since they did not inspire admiration but a kind of pity.

Samper claimed he detected ethno-national traits (rather than his own bias) when he judged other travelers' behaviors as alcoholic, arrogant, or narrow-minded. For instance, he observed a loud quarrel between a "bourgeois" French couple during a long journey to Paris. From this brief study, he concluded that French men were too "elastic by temperament," changing to fit the "spirit" of each country they visited. In contrast, French women were too rigid. He judged them coquettish, flippant, indifferent, independent, seductive, spiritualist, contradictory, and obsessed with fashion.[16] He criticized illiberal thinkers, too: fundamentalist Muslims had "destroy[ed] the notion of liberty and responsibility, and establish[ed] the slavery and degradation of women!" Similarly, Christian "hypocrites" persecuted Islam "without renouncing their own fatalism or having done anything for the liberty and dignity of women."[17] Somewhere between these two extremes of flexibility and rigidity lay the middle ground that constituted this Colombian's notion of the ideal national character.

In criticizing other people and cultures, Samper revealed his own limits. He was learning to be a spouse, father, son-in-law, and Colombian gentleman while traveling abroad. Over his first year and a half of reporting from England, Spain, and France, he narrated his experiences as *un colombiano*. Why did Samper's first volume of travels present him as a lone traveler? From today's vantage point, it seems remarkably self-centered that an author could travel thousands of miles by boat, rail, and carriage with a spouse, mother-in-law, and young daughters in tow and not mention them in the memoir's first sentences. His authorial identity was firmly rooted in nineteenth-century conventions of selfhood. Marriage and parenthood had negligible effect on his confident "universal" (elite male) voice. The first volume of his travelogue expressed a national identity in which women and children did not figure.

Where were his wife, mother-in-law, and children during his first travels, which made up the essays for the first volume of *Viajes*? During his first trips around Spain and the Mediterranean in 1858,

Acosta and her mother stayed "home," setting up temporary quarters in Paris and caring for the infant Carolina and the toddler Bertilda. The difficulties of travel in certain regions may have figured in their decision. Spain, for example, lacked an extensive railway at this point. In 1848 only one thirty-three-kilometer railroad connected Barcelona and Mataró; in 1851 a new line linked Madrid and Aranjuez. Only after the family returned to South America in 1862 did a rail line traverse Spain from border to border (1862 to Portugal, 1863 to France). Moving overland in many regions of Europe required travel by foot, on horseback, or in a horse-drawn carriage on roads of variable quality. By midcentury there were steamships plying the oceans and inland waterways, but sail- and rowboats were still more common. Besides, all travel posed risks.

Beyond the logistics and uncertainties of travel, it was likely Acosta's own responsibilities as a working mother that lay behind her decision not to join her husband in his first travels as a correspondent. At twenty-six years of age, she wrote under a variety of pseudonyms for the literary journal *El Mosaico*, which later merged with the *Biblioteca de Señoritas*. It was one of the first periodicals ever directed at Latin American women, and it reached a significant number of readers in Colombia, Ecuador, and Venezuela. In January 1859 Acosta began a popular bimonthly report on life in Paris ("Revista Parisiense") in which she chronicled developments in literature, the arts, and the sciences, as well as fashion trends, theater, parties, and seasonal events. She temporarily suspended that work in June 1859, since the leisured classes had fled the city's oppressive summer heat.[18] She and Samper did the same. As they embarked for Switzerland, their two young daughters stayed in Paris with their grandmother.

In October 1859 Acosta began publishing installments of "Recuerdos de Suiza," under the pseudonym "Andina" in *El Mosaico* and *El Comercio*, detailing their monthlong trip to and through central Switzerland.[19] Interestingly, her Swiss essays referred to "us" and "we" without specifying Samper. Acosta seemed to expect her readers to recognize them as a couple. Indeed, even her pen names were so famous that readers could indeed do so. One of her contemporaries, the Peruvian author Mercedes Cabello de Carbonera (1845–1909), remarked that "the newlyweds transplanted themselves from Colombia to Paris, from where they sent considerable correspondence; the pen and wit of [Acosta] were the most important factor that lent them their pleasantness and brightness; those of her essays published in *El Comercio*, which had a large circulation in Lima then as now, were read avidly and held in very high esteem by those who could assess the merits of those two literary talents."[20] While Samper may have held political and professional offices, it seems that many readers preferred Acosta's writing to his. She became Colombia's most prolific author, writing everything from poetry and novels to journalism and history.

Swiss Alpine Tourism

July 16, 1859, dawned hot and humid. Samper and Acosta left Paris by an early train heading for Geneva, Switzerland, where they arrived just before midnight. On July 19 a gigantic carriage, drawn by five horses and seating twenty-six passengers, took them to Chamonix, France, where they hired a guide and mules to ride and hike around the Montanvers glacier. From there, at 5:00 a.m. on July 21, they rode the same "enormous and prudent" mules for a full day (about sixty miles) to the shore of Lake Geneva. A half-hour steamboat ride brought them into Switzerland again, to the beautiful "city in miniature" of Vevey.[21] Over the next week they visited the cities of Lausanne, Neuchâtel, Fribourg, and Bern by stagecoach. After an hour-long train trip from Bern, they marveled at the "astonishingly beautiful Lake Thun" before making a visit on July 26 to the town of Interlaken, at the foot of the Oberland Alps in central Switzerland.[22]

In early August the couple left to explore the natural spectacles near Grindelwald, from ice caverns to the mighty peaks of Eiger and Jungfrau. They traveled next to Lake Brienz, taking a short steamboat ride and climbing for several hours around the Giessbach waterfalls. They hired horses and rode over Brünig Pass and down to the "poor hamlet" of Lungern, where they spent a miserable night on bad mattresses in a stiflingly hot room. They fled

3. Victorian hikers on the Chamonix glacier in the Savoy Alps, France, 1867. Thomas Cook Archives.

at 3:00 a.m. by carriage to Sarnen, the canton capital, then took a steamboat to Lucerne.[23] After enjoying the city and Lake of the Four Cantons for a few days, they crossed to the opposite shore by steamboat and then traveled by coach toward Rigi on August 4, 1859. They hiked in the Küssnacht am Rigi area for a few days then journeyed by carriage and rail to tour Zurich, Schaffhousen, and Basel. They left Switzerland on August 14, 1859.

Samper and Acosta's Swiss Alps circuit was firmly situated in the European tourist market. Alpine infrastructure had grown with the hospitality industry. Investors built hotels, train stations, and lake and river ports. Canton governments dug tunnels, built roads, and laid railway track. Guides, drivers, baggage handlers, cooks, bakers, waiters, and hotel staff set to their tasks with increased fervor in summer. One of the first Alpine hotels opened in Rigi in 1817, making it the most visited mountain in Switzerland by midcentury. Elite demands for Alpine recreation pushed the development of luxury lakeshore accommodations like Hotel Schweizerhofen in Lucerne and Hotel Baur au Lac in Zurich.[24] The Alps were a place to explore a European culture of leisure in shared spaces like hotels, restaurants, and hiking trails.[25]

4. Adolphe Braun, *Swiss Alps above Chamonix*, ca. 1858. Harvard University.

The Alps developed as an offshoot of the Grand Tour for adventurers and Romantics. In 1786 the Alps opened to a new kind of tourism—mountaineering—when two local hunters first ascended Mont Blanc, the highest peak in Europe, and showed the route to Genevan scientist Horace de Saussure. British interest in Alpine mountaineering exploded after journalists began reporting on this ascent in the 1850s. The press celebrated climbers as imperial heroes, creating such a popular and aspirational image for the middle class that a decade later twenty or thirty summited Mont Blanc each year.[26]

The particular hassles, costs, and dangers of travel to the Alps also defined its prestige for international visitors. While some Europeans wanted to climb mountains for athletic or competitive reasons, others cued into early nineteenth-century Romanticism and trekked the mountains for more ineffable reasons. Romanticism in arts and literature emphasized the unpredictable, uncontrollable power of nature. Romantics sought individual, subjective experiences, often through human struggles with nature's forces.[27] The Alps provided Romantics with the perfect recipe for beauty and terror that would conjure the sublime.

How did Acosta and Samper understand and express their national, gender, and class identities in a leisure culture dominated by British and German tourists and writers? In a simple sense

Acosta lamented the multitude of German and English tourists. About their hotel in Rigi, Acosta wrote, "All races of the civilized world are seen there . . . everything is spoken about and in every language, except our own, which leaves us the honor of being the only representatives from all of Latin America."[28] But her distance from home also revealed her deeper belonging to Colombia and to a broadly Latin American identity. Acosta felt keenly her foreignness and class position while traveling among middle-class aspirants to the Swiss Alpine "club." She noted the cosmopolitanism of some Alpine towns and observed the isolation and poverty of others. Readers could infer the comparison to certain Latin American regions. As a privileged Colombian, Acosta possessed biases and sympathies that situated her awkwardly between the Old World aristocracy and the New World bourgeoisie.

Acosta carped at the English middle classes who had sullied the aristocratic timbre of the Grand Tour, which she clearly admired in certain respects. Her hostility toward English tourists stemmed from her association of the English with industrialized technology like trains and the artifice of civilization. For instance, she praised Lake Neuchâtel for its natural beauty, which was enhanced by "having few English tourists plowing through its crystalline green waves and commandeering the best spots in steamships, coaches, and hotels."[29] At Interlaken, Acosta bemoaned the presence of the English, who "had invaded that charming place, and everywhere appeared the monumental crinolines of the ladies." After the English women passed by in their huge hoop skirts, Parisian women would laugh and rearrange the long trains of their own dresses. "Men spend their lives in the casino and women spend it warring against each other," Acosta sighed, adding, "Todo el mundo es Popayán!" In other words, the world is much the same everywhere.[30]

Quiet, natural places magnified Acosta's perception of England's multitudes and mistakes. She particularly resented their behavior at historical sites and in public transportation. Acosta developed a theory about English tourists within the first days of their Swiss trip. After being squashed between a "serious and arrogant" English family in a gigantic twenty-six-seat carriage they boarded in

Zurich, she explained her stereotyping. The first "perfectly harmless" English type simply followed his doctor's orders to travel: the "rich and irritable Englishman who tries to cure his organic and hereditary *espleen* [spleen] going around the world."[31] Coldly serious, the wealthy never tried to conserve memories of their travels in their minds; instead, they recorded them in portfolios. She remarked upon this English type a few weeks later, while visiting a historical site in Lucerne that memorialized the Swiss Guards killed defending France's King Louis XVI during the radicalized revolution in August 1792. At the site, "in exchange for a small tip," a man in an old Swiss Guard uniform told of his compatriots' deaths "with theatrical accent," swearing he had witnessed these as a young drummer boy. "But since the Swiss are wise enough to exploit a traveler's credulity," Acosta sagely cautioned her readers, "we did not put much faith in the old man's story." However, Acosta noted, "an Englishwoman listened to him admiringly, interrupting him from time to time with a prolonged *shocking*, while her husband took careful notes in his notebook."[32] She cued the humor of the English upper-class pretense with this anecdote, at the same time offering a lesson in tourist savvy to her South American readers.

As bad as the wealthy may have been, Acosta still lamented that they were "changing or ending" and being replaced by the insufferable English middle classes. She condemned them as "not only not harmless, but if they had been invented in the time of Moses, they would have been sent to Egypt as the eighth and most terrible plague [of locusts]." They bargained for deals because they obsessed over prices, which thinly disguised an ugly obsession with lucre: "Since they are not rich, or they do not want to spend money, they travel economically, annoying anyone who comes near them with their unprecedented demands; they are abrupt and extremely spoiled." Middle-class tourists "think it is indispensable to bring home as many souvenirs as possible," as proof of their travels. They embodied hypocrisy: "Being poor, and not being able to buy them, they do not have the slightest scruple in stealing what most strikes them in museums, cutting slices from unguarded furniture, statues, and decorations."[33] Acosta vilified their petit bourgeois appetite for arti-

5. Unidentified artist, *"The English have invaded."* From Soledad Acosta de Samper's collage in her printed journal of Swiss travels. National Library of Colombia.

facts. Nonetheless, she found the "mortal hatred" between these two classes "odd" and "terrible." English tourists, whether stodgy or sinful, undermined the Alpine sublime. Acosta resigned herself, grumpily, to the unstoppable socioeconomic changes that had increased with European travelers' access to the Alps. Hotel guestbooks reveal some of the new patterns of economics and tourist culture in the Alps.[34]

By midcentury Alpine guestbooks had become objects of curiosity in their own right. A "Book of Travelers," Acosta noted, was "found in any famous site in Switzerland." The simplest function of the registers was to record guests' names and nations. However, British visitors to the Alps were famous for adding their observations and commentary to these registers. They criticized or praised hotel management, staff, other guests, food, prices, guides, and routes; some even listed challenges for other adventurers.[35] Guestbooks encouraged imitation by those who could read the most common languages represented (English, French, and German). Reading these comments, tourists might plan itineraries, shape expectations of a hike or a site, share in the common experiences of travelers, or laugh at the foibles of others. Acosta paged through one Montanvers guestbook while waiting for their mules to be readied just after their glacier hike. "The most characteristic entries," she sniffed, are from "those who travel solely so they can say that they had been to this or that place." She transcribed two typical cases: an Englishman listed the price of lunch for his family, and a U.S. tourist wrote that he arrived on foot, with a bad stomachache, on U.S. Independence Day. Feeling better after a brandy, he wrote, "Long live the State of Massachusetts!"[36] Acosta used the inadvertent humor of such visitors—who focused on their guts and ignored the glory around them—to illustrate what not to do as an educated traveler in the Alps.

Acosta's criticisms of other travelers implied that she was not only more sensitive but also more sensible. She often remarked on the absurd fastidiousness of British manners, though the French drew her fire too. For example, at the start of their trip in July, Acosta and Samper shared a sweltering train car from Paris to Geneva with two wary English women and their brother. The women asked for tea at every stop, each time dismayed that no one ever had it to sell, and they sighed longingly for an English "canteen." After overcoming the women's mistrust with her English fluency, Acosta tried to persuade them to shed their voluminous hats and capes, as she had done in the terrible heat. "They refused the advice with apparent horror," and only then did Acosta realize that the long curls of hair hanging down their cheeks were actually *attached* to their hats. Though "suffocating and envious," they were "unwilling to lose their British dignity." Their unswerving dedication to English manners smacked of vanity, artifice, and silliness but also the height of ignorance and disregard for the culture and climate in which they traveled.[37] She illustrated to her South American readers the line between impractical adherence to cultural norms and pragmatic flexibility. Travelers should not have to renounce their core values, but they would have to adjust some habits and expectations.

Acosta saw herself as different from other Alpine tourists, pushing at some gender and class norms but not dispensing with all of them. Women rode mules and walked well-established, lower-altitude trails while wearing crinolines, and often carrying parasols, as early photographs of the Alps show.[38] Likewise, both Acosta and Samper rode mules over difficult terrain. However, Acosta judged certain fashions or transport to be excessively passive or decorative given the rugged landscape and altitude. For instance, she lamented that most English and Germans viewed the peaks from their hotel balconies rather than from hiking trails. Acosta was certainly not so timid a sightseer as this type of tourist, one content to view the landscape in paintings or through windows. She was a rare woman who exerted herself by hiking the more difficult Alpine trails and glaciers. But neither was she as intrepid as Lucy Walker, the British mountaineer then breaking European gender conventions and scaling incredible heights.[39] Acosta did not describe how she dressed for her hikes, but she did criticize those who adhered to fashion over practicality. After one long hike, for example, she encountered several solemn, well-clad English men riding "on mules saddled with women's light saddles." Later she observed a *petrimetre* (roughly, "French dandy") riding a carriage horse with bells. He was dressed entirely in white, wearing a monocle, a straw hat with blue veil, and carrying a thin cane. "This delicate character" was headed toward the mountains and "whistling a piece of opera."[40] She objected not to the animals that were providing assistance to travelers but rather to their (and his) decoration. As ever in her Alpine essays, Acosta enjoyed relating the absurd clash between the everyday and the ethereal.

Samper praised his wife for this quality: she did not just observe nature but moved in it to find new experiences. He wrote, "My wife, beyond having a decided taste for the beautiful spectacles of nature, is animated on trips, having had the whim to propose that we walk down from the Rigi on foot, in the direction of Küssnacht. Even though the descent is rather exhausting and requires almost three hours, we did not renounce the project."[41] Acosta's physical stamina, mental acuity, and maternal nature became for him material evidence of a beloved national home, not just a domestic one.[42] Samper delighted in his wife's companionship for that reason. About an excursion to the Grindelwald glacier and waterfalls, he wrote, "I confess that I have never in my life experienced such supreme happiness as in these moments. I had before and behind me all the beauty of a severe and savage nature, grandiloquent for its rumors and aspect, dazzling and pregnant with infinite poetry; And at my side, dreaming like me, my beloved wife, the companion and guardian angel of my life, we only felt the adoration of the beautiful and the sublime."[43] Here Acosta reflected on the "strange mystery" of the "blue glass" cavern and the nearby Lütschine River that surged from the glacier "like a captive who runs away, shaking his chains, and dragging along the rubble of his prison, in the form of stones and enormous masses of ice that swim, swirling, over its ashen bosom."[44] Both writers wanted their readers to see the importance of active reflection on natural beauty. Their moving meditations freed them to examine the world and their position in it. For Samper, this set them apart from the "pseudo-travelers" he so disliked.[45] Reading these observations, their South American readers could better appreciate their own lands, cultures, and people, even if they never took a Grand Tour of Europe.

The weeks of arduous journeying and sustained reflection also changed Samper's perspective on both the Alps and his wife. In this incredible terrain he could better appreciate Acosta's desirability as a wife and companion. The second volume of *Viajes*, unlike his earlier volume of solitary travels, noted his wife's active and important presence as a traveling partner. Even his grammar became more inclusive (for example, "we chose" rather than "I chose"). A footnote explained the obvious change in his narrative: "When I speak in plural, I refer to my wife and myself."[46] Traveling without their daughters, Samper and Acosta drew closer, both as a couple and as Colombians. She embodied his ideal virtues of activity, durability, flexibility, humility, prudence, sobriety, and religiosity. Samper reminded his South American readers to maintain their national gender traits, as Acosta did, but also to study other languages, as she had. He was especially pleased that "an eminent [Swiss] chemist" had a lively conversation with Acosta and that, after "conversing in English with [a U.S. railway magnate], my wife influenced him much in our favor."[47] For Samper, she was the model wife of a citizen of a young South American republic who could hold her own in this "enlightened" European milieu. In this, Samper had grown to recognize that his wife opened just as many doors into society and politics for him as he did for her.

For Acosta, the Alps evoked the Andes: home. Her abiding connection to Colombia grew more vivid and beloved as they traveled. When they set off for Switzerland, Samper remembered her saying, "Finally, we are going to visit that country of mountains and lakes, the father of almost all the great rivers of the European continent. That will produce emotions in us that will evoke at every moment the beloved image of home."[48] Acosta's image of a fatherly Switzerland could have implied that their home was the opposite: maternal, wifely, and emotional. Instead, she saw the affinity between the Alps and the Andes. Peoples of both mountain regions insisted on independence from European cultural dominance. For their readers, that the Colombians had even reached the Alpine circuit demonstrated their tenacity, flexibility, and education. If they could do it, why could not other equally privileged South Americans? Acosta's reports in particular taught her readers how to travel as a respectful tourist: take up little space, speak softly, don't talk about money, and keep memories of, not objects from, your travels.

Samper and Acosta turned the critical lens of the travel genre onto the Swiss Alps and the tourists who flocked there. The pair placed themselves next to Europeans, moved among them as equals, and reported their observations through a combination of humorous description, practical explanation, and stiff critique. Taken together, theirs was a missive to future South American tourists: part invitation, part caution.

Notes

1. Carrera, "Escritura femenina y literatura de viajes," esp. 114–15. See also Frawley, *Wider Range*.

2. Nineteenth-century U.S. travelers to Europe, such as Ralph Waldo Emerson and Mark Twain, are often studied; see Melton, *Mark Twain, Travel Books,*

and Tourism. For studies of European and U.S. travelers to Latin America, see Denegri, *"Desde la Ventana"*; and Echenberg, *Humboldt's Mexico.* Scholars also analyze gender, race, and nation in travel literature; see Mulvey, *Transatlantic Manners*; and Hill, *Britain and the Narration of Travel.*

3. See Fey and Racine, *Strange Pilgrimages*; Sznajder and Roniger, *Politics of Exile in Latin America*; Sznajder and Roniger, "Political Exile in Latin America"; Graham-Yooll, "Wild Oats They Sowed"; Kaminsky, *After Exile*; and Kaplan, *Questions of Travel.*

4. Pratt, *Imperial Eyes.*

5. Samper, *Historia de un alma*, 354–55.

6. See "Cronología" in Acosta de Samper, *Novelas y cuadros de la vida suramericana*, 403–6; and Hinds, "José María Samper," 165.

7. *El Mosaico: Periódico de la juventud, destinado exclusivamente a la juventud* (1858–60, 1862–65) began publishing *Biblioteca de Señoritas* (July–October 1859) after its independent distribution in thirty-four Colombian cities, Caracas, Venezuela and Quito, Ecuador. Acosta used pseudonyms (Aldebarán, Andina, Bertilda, Renato) until returning to South America in 1869, when she published her first signed book. Unless otherwise noted, all translations from Acosta's and Samper's works are mine.

8. Samper, *Viajes*, 2:1.

9. Samper, *Viajes*, 2:2–3.

10. Samper, *Viajes*, 1:13–14.

11. Several important works have examined the ideas of race, nation, and ethnicity in the writings of Samper and other nineteenth-century Colombian elites. See Gobat, "Invention of Latin America"; Urueña, "La idea de heterogeneidad racial"; Rojas de Ferro, "Identity Formation, Violence, and the Nation-State"; Aponte Ramos, "El cuerpo étnico"; Trigo, *Subjects of Crisis*; D'Allemand, "Quimeras, contradicciones, ambigüedades"; McGuinness, *Path of Empire*; and Safford, "Race, Integration, and Progress."

12. Samper, *Viajes*, 1:2.

13. Samper, *Viajes*, 1:12.

14. Samper, *Viajes*, 1:194.

15. Samper, *Viajes*, 1:195.

16. Samper, *Viajes*, 1:138–39.

17. Samper, *Viajes*, 1:366.

18. Alzate, *Soledad Acosta de Samper*, 80–82, 86–87.

19. Soledad Acosta de Samper's "Viajes: Recuerdos de Suiza" were first published in *El Mosaico*, no. 41 (1859): 325–28 and no. 46 (1859): 365–68, which is not available digitally. She reprinted these in installments in *La Mujer* 3 (1879–80) under the byline S. A. de S., which this chapter cites. See Banco de República Colombia Biblioteca Virtual, http://babel.banrepcultural.org/cdm/ref/collection/p17054coll26/id/1661.

20. Cabello de Carbonera quoted in Ordóñez, *De voces y de amores*, 46.

21. Acosta, "Viajes" *La Mujer* 3, no. 30 (December 15, 1879): 133, 135–37.

22. Acosta, "Viajes," *La Mujer* 3, no. 31 (February 1, 1880): 159–65. She remarked on the speed of train travel: "In another time, now many years ago, it took a whole day to go from Bern to Thun; later it took four hours and now it takes one hour by rail" (163).

23. Acosta, "Viajes," *La Mujer* 3, no. 32 (February 15, 1880): 183–89.

24. For more on British tourism in Switzerland, see Tissot, "How Did the British Conquer Switzerland?," 21–54; and Ring, *How the English Made the Alps*.

25. Ray Oldenburg's "third place" concept distinguished home and work from sites of voluntary, informal, and regular sociability such as cafés, pubs, and salons. Oldenburg, *Great Good Place*.

26. Peter Hansen argues that middle-class British men, concerned that theirs was becoming "a wealthy but unmanly society," used mountaineering to construct "an assertive masculinity." Even Edward Whymper's 1865 Matterhorn tragedy (four men died on descent) did not deter these climbers. Hansen, "Albert Smith, the Alpine Club, and the Invention of Mountaineering," 303–4.

27. Edmund Burke defined the sublime as emerging from "whatever is fitted in any sort to excite the ideas of pain, and danger . . . whatever is in any sort terrible . . . or analogous to terror . . . the strongest emotion which the mind is capable of feeling." E. Burke, *Philosophical Enquiry into the Origin of our Ideas of the Sublime and Beautiful*, 13.

28. Quoted in Vallejo, "La perspectiva femenina de un viaje," 98, 104.

29. Acosta, "Viajes," *La Mujer* 3, no. 31 (February 1, 1880): 159.

30. Acosta, "Viajes," *La Mujer* 3, no. 31 (February 1, 1880): 165.

31. Acosta, "Viajes," *La Mujer* 3, no. 30 (December 15, 1879): 133. She referred to ancient Greek humoral theory that the spleen dictated emotions or temperament. Literary references to the small blood-filtering organ abound; for example, the irritable Cassius in Shakespeare's *Julius Caesar*, 4.3: "By the gods, you shall digest the venom of your spleen, though it do split you." By Acosta's time the spleen also was commonly associated with melancholy and excessive laughter.

32. Acosta, "Viajes," *La Mujer* 3, no. 33 (March 1, 1880): 210.

33. Quoted in Vallejo, "La perspectiva femenina," 98.

34. Scholars have found that the names of British travelers dominated Alpine hotel guestbooks and registers. See Michalkiewicz and Vincent, "Victorians in the Alps," 75–90. It was the British who had created the genre of hotel guestbooks and visitor registers; see James, "'British Social Institution.'"

35. Michalkiewicz and Vincent, "Victorians in the Alps," 84–86.

36. Acosta, "Viajes," *La Mujer* 3, no. 30 (December 15, 1879): 135. No trace of Samper and Acosta has yet been found in archived guestbooks. Patrick Vincent, email to author, April 19, 2017.

37. Acosta, "Viajes," *La Mujer* 3, no. 29 (December 1, 1879): 109–10.

38. In a hotel guestbook, a visitor had drawn a picture of a woman riding a mule on the long ascent to Gornergrat. Another noted the rarity of a woman on

a high-altitude trek: "On Saturday Aug. 12 this party visited Weisshorn returning over the Stockhorn, Hochthaligrat, Gornergrat ... to the Riffelhouse. Mrs. Tower is the first lady who has ever been at the Weisshorn or on the Stockhorn or over the rough passage between the latter at the Hochthaligrat. ... Guide Jean Baptiste Brantscher is highly recommendable for every kind of excursion in this vicinity." Quoted in Michalkiewicz and Vincent, "Victorians in the Alps," 82–83.

39. In 1864 Lucy Walker (1836–1916) gained fame when she climbed the Eiger (3,970 meters, or 13,000 feet), one of the most terrifying peaks in the Alps. She wore the same cumbersome skirts as most Victorian women, despite the danger these may have added to her climb.

40. Acosta, "Viajes," *La Mujer* 3, no. 32 (February 15, 1880): 184.

41. Samper, *Viajes*, 2:216.

42. Acosta, "Viajes," *La Mujer* 3, no. 34 (March 24, 1880): 234. Acosta was interested in the legend that Pontius Pilate was buried on Rigi, an interest that arose from her strong Catholic faith. Vallejo, "La perspectiva femenina," 105. See also Vallejo Mejía, "Soledad Acosta de Samper y el periodismo como cátedra de moral"; and Plata, "Soledad Acosta de Samper."

43. Samper, *Viajes*, 2:159–60.

44. Acosta, "Viajes," *La Mujer* 3, no. 32 (February 15, 1880): 184.

45. Samper, *Viajes*, 2:1.

46. Samper, *Viajes*, 2:65.

47. Samper, *Viajes*, 2:290–95, 325.

48. Samper, *Viajes*, 2:8.

Bibliography

Acosta de Samper, Soledad. *Biblioteca de Señoritas* (July–October 1859). Hemeroteca Digital Histórica, Red Cultural del Banco de la República (Bogotá, Colombia). http://babel.banrepcultural.org/cdm/ref/collection/p17054coll26/id/3351.

———. *Diario íntimo y otros escritos*. Edited by Carolina Alzate. Bogotá: Alcaldía Mayor de Bogotá, Insituto Distrital Cultura y Turismo, 2004.

———. *El Mosaico: Periódico de la juventud, destinado exclusivamente a la juventud* (1858–60, 1862–65). Hemeroteca Digital Histórica, Red Cultural del Banco de la República (Bogotá, Colombia). http://babel.banrepcultural.org/cdm/ref/collection/p17054coll26/id/2757.

———. *Novelas y cuadros de la vida suramericana*. Edited by Montserrat Ordóñez. Ghent, Belgium: Eug. Vanderhaeghen, 1869; Bogotá: Edición Uniandes, 2004.

Acosta de Samper, Soledad. "Viajes: Recuerdos de Suiza." First published in *El Mosaico* (1859) no. 41: 325–28 and no. 46: 365–68. Reprinted in *La Mujer* 3 (1879–80), no. 29: 109–12, no. 30: 133–37, no. 31: 159–65, no. 32: 183–89, no. 33: 210–12, no. 34: 233–34, no. 35: 256–59, no. 36: 280–83. Accessed via Hemeroteca Digital Histórica, Red Cultural del Banco de la República

(Bogotá, Colombia). http://babel.banrepcultural.org/cdm/ref/collection/p17054coll26/id/1661.

Agosín, Marjorie, and Julie H. Levison. *Magical Sites: Women Travelers in 19th Century Latin America*. Buffalo NY: White Pine Press, 1999.

Alzate, Carolina. "El diario epistolar de dos amantes del siglo XIX: Soledad Acosta y José María Samper." *Revista de Estudios Sociales* (Bogotá, Colombia) 24 (May–August 2006): 33–37.

———. *Soledad Acosta de Samper y el discurso letrado de género, 1853–1881*. Madrid: Iberoamericana, Vervuert, 2015.

Alzate, Carolina, and Isabel Corpas, eds. *Voces diversas: Nuevas lecturas de Soledad Acosta de Samper*. Bogotá: Instituto Caro y Cuervo, Universidad de los Andes, 2016.

Aponte Ramos, Lola. "El cuerpo étnico en la constitución del espacio urbano y el proyecto nacional de José María Samper." In *Chambacú, la historia la escribes tú*, edited by Lucía Ortiz, 349–60. Madrid: Iberoamericana, 2007.

Arango Rodríguez, Selen Catalina. "De la poesía a la prosa, o imágenes de la formación femenina en el proyecto de nación colombiano del siglo XIX." *Historia Caribe* (Barranquilla) 17 (2010): 67–88.

Aristizábal, Magnolia. *Madre y esposa: Silencio y virtud; Ideal de formación de las mujeres en la provincia de Bogotá, 1848–1868*. Bogotá: Universidad Pedagógica Nacional, 2007.

Bassnett, Susan, ed. *Knives and Angels: Women Writers in Latin America*. London: Zed Books, 1990.

Burke, Edmund. *A Philosophical Enquiry into the Origin of Our Ideas of the Sublime and Beautiful*. London, 1757. Accessed at https://quod.lib.umich.edu/e/ecco/004807802.0001.000/1:4.7?rgn=div2;view=fulltext.

Burke, Janet, and Ted Humphrey, eds. *Nineteenth-Century Nation Building and the Latin American Intellectual Tradition: A Reader*. Indianapolis: Hackett, 2007.

Carrera, Elena. "Escritura femenina y literatura de viajes: Viajeras inglesas en la España del XIX, lugares comunes y visiones particulares." In *Diez estudios sobre literatura de viajes*, edited by Manuel Lucena Giraldo and Juan Pimentel, 109–30. Madrid: Consejo Superior de Investigaciones Científicas, Instituto de la Lengua Española, 2006.

D'Allemand, Patricia. "Quimeras, contradicciones, ambigüedades en la ideología criolla del mestizaje: El caso de José María Samper." *Revista de Historia y Sociedad* (Medellín) 13 (November 2007): 45–63.

Denegri, Francesca. "*Desde la Ventana*: Women 'Pilgrims' in Nineteenth-Century Latin-American Travel Literature." *Modern Language Review* 92, no. 2 (1997): 348–62.

Dueñas Vargas, Guiomar. *Of Love and Other Passions: Elites, Politics, and Family in Bogotá, Colombia*. Albuquerque: University of New Mexico Press, 2015.

Echenberg, Myron. *Humboldt's Mexico: In the Footsteps of the Illustrious German Scientific Traveler*. Montreal: McGill-Queen's University Press, 2017.

Fey, Ingrid E., and Karen Racine, eds. *Strange Pilgrimages: Exile, Travel, and National Identity in Latin America, 1800s–1990s.* Wilmington DE: SR Books, 2000.

Frawley, Maria. *A Wider Range: Travel Writing by Women in Victorian England.* Rutherford NJ: Fairleigh Dickinson University Press, 1994.

Galeana, Patricia, coord. *Latinoamérica en la conciencia europea: Europa en la conciencia latinoamericana.* México DF: Archivo General de la Nación, 1999.

Gobat, Michel. "The Invention of Latin America: A Transnational History of Anti-Imperialism, Democracy, and Race." *American Historical Review* 118, no. 5 (2013): 1345–75.

Gómez Ocampo, Gilberto. "Soledad Acosta de Samper, 1833–1913, Colombian prose writer." In *Encyclopedia of Latin American Literature*, edited by Verity Smith, 1–3. Chicago: Fitzroy Dearborn, 1997.

Graham-Yooll, Andrew. "The Wild Oats They Sowed: Latin American Exiles in Europe." *Third World Quarterly* 9, no. 1 (1987): 246–53.

Hansen, Peter. "Albert Smith, the Alpine Club, and the Invention of Mountaineering in Mid-Victorian Britain." *Journal of British Studies* 34, no. 3 (1995): 300–324.

Hill, Kate, ed. *Britain and the Narration of Travel in the Nineteenth Century: Texts, Images, Objects.* Burlington VT: Ashgate, 2016.

Hincapié, Luz M. "Amor, matrimonio y educación: Lecturas para mujeres colombianas del siglo XIX." *Credencial Historia*, no. 277 (January 2013). https://www.banrepcultural.org/blaavirtual/revistas/credencial/enero-2013/amor-matrimonio-y-educacion.

Hinds, Harold E. "José María Samper: The Thought of a Nineteenth-Century New Granadan during His Radical-Liberal Years, 1845–1865." PhD diss., Vanderbilt University, 1976.

James, Kevin. "'[A] British Social Institution': The Visitors' Book and Hotel Culture in Victorian England and Ireland." *Journeys: The International Journal of Travel Writing* 13, no. 1 (June 2012): 42–69.

Kaminsky, Amy K. *After Exile: Writing the Latin American Diaspora.* Minneapolis: University of Minnesota Press, 1999.

Kaplan, Caren. *Questions of Travel: Postmodern Discourses of Displacement.* Durham NC: Duke University Press, 1996.

Kerber, Linda K. "The Republican Mother: Women and the Enlightenment—An American Perspective." *American Quarterly* 28, no. 2 (1976): 187–205.

McGuinness, Aims. *Path of Empire: Panama and the California Gold Rush.* Ithaca NY: Cornell University Press, 2008.

McMahon, Lucia. *Mere Equals: The Paradox of Educated Women in the Early American Republic.* Ithaca NY: Cornell University Press, 2012.

Melton, Jeffrey Alan. *Mark Twain, Travel Books, and Tourism: The Tide of a Great Popular Movement.* Tuscaloosa: University of Alabama Press, 2002.

Mesa Gancedo, Daniel. "Lecturas cruzadas y la escritura del diario: Soledad Acosta." *Decimonónica* 5, no. 2 (2008): 1–32.

Meyer, Doris, ed. *Rereading the Spanish American Essay: Translations of 19th and 20th Century Women's Essays*. Austin: University of Texas Press, 1995.

Michalkiewicz, Katarzyna, and Patrick Vincent. "Victorians in the Alps: A Case Study of Zermatt's Hotel Guest Books and Registers." In *Britain and the Narration of Travel in the Nineteenth Century: Texts, Images, Objects*, edited by Kate Hill, 75–90. New York: Routledge, 2016.

Mulvey, Christopher. *Transatlantic Manners: Social Patterns in Nineteenth Century Anglo-American Travel Literature*. New York: Cambridge University Press, 1990.

Oldenburg, Ray. *The Great Good Place: Cafés, Coffee Shops, Community Centers, Beauty Parlors, General Stores, Bars, Hangouts, and How They Get You through the Day*. 2nd ed. New York: Marlowe, 1997.

Ordóñez, Monserrat. *De voces y de amores: Ensayos de literatura latinoamericana y otras variaciones*. Edited by Carolina Alzate et al. Bogotá: Grupo Editorial Norma, 2005.

———. "One Hundred Years of Unread Writing: Soledad Acosta, Elisa Mújica, and Marvel Moreno." In *Knives and Angels: Women Writers in Latin America*, edited by Susan Bassnett, 132–44. London: Zed Books, 1990.

Ordóñez, Montserrat, and Carolina Alzate, eds. *Soledad Acosta de Samper: Escritura, género y nación en el siglo XIX*. Madrid: Iberoamericana, 2005.

Plata, William Elvis. "Soledad Acosta de Samper: La modernidad en el catolicismo." In Alzate and Corpas, *Voces diversas*, 391–416.

Pratt, Mary Louise. *Imperial Eyes: Travel Writing and Transculturation*. London: Routledge, 1992.

———. "Soledad Acosta de Samper, 1833–1913." In Meyer, *Rereading the Spanish American Essay*, 71–76.

Ring, Jim. *How the English Made the Alps*. London: John Murray, 2000.

Rojas de Ferro, Cristina. "Identity Formation, Violence, and the Nation-State in Nineteenth-Century Colombia." *Alternatives: Global, Local, Political* 20, no. 2 (April–June 1995): 195–224.

Roniger, Luis, James N. Green, and Pablo Yankelevich, eds. *Exile and the Politics of Exclusion in the Americas*. Portland OR: Sussex Academic Press, 2014.

Safford, Frank. "Race, Integration, and Progress: Elite Attitudes and the Indian in Colombia, 1750–1870." *Hispanic American Historical Review* 71, no. 1 (1991): 1–33.

Samper, José María. *Historia de un alma: Memorias íntimas y de historia contemporánea*. Bogotá: Imprenta de Zalamea Hermanos, 1881. https://archive.org/download/historiadeunaal00sampgoog.

———. *Viajes de un colombiano en Europa*. 2 vols. Paris: Imprenta de E. Thunot y Compañía, 1862. Vol. 1 available at http://www.gutenberg.org/ebooks/14329; vol. 2, http://www.gutenberg.org/ebooks/15054.

Stepan, Nancy Leys. *Picturing Tropical Nature*. Ithaca NY: Cornell University Press, 2001.

Sznajder, Mario, and Luis Roniger. "Political Exile in Latin America." *Latin American Perspectives* 34, no. 4 (2007): 7–30.

———. *The Politics of Exile in Latin America*. New York: Cambridge University Press, 2009.

Tissot, Laurence. "How Did the British Conquer Switzerland? Guidebooks, Railways, Travel Agencies: 1850–1914." *Journal of Transport History* 16, no. 1 (1995): 21–54.

Trigo, Benigno. *Subjects of Crisis: Race and Gender as Disease in Latin America*. Middletown CT: Wesleyan University Press; Hanover NH: University Press of New England, 2000.

Urueña, Jaime. "La idea de heterogeneidad racial en el pensamiento político colombiano: Una mirada histórica." *Análisis Político* (Bogotá) 22 (May–August 1994): 5–25.

Vallejo, Catherina. "La perspectiva femenina de un viaje: 'Recuerdos de Suiza' de Soledad Acosta de Samper (1859) y *Viajes de un colombiano en Europa* (1862) de José María Samper." In Alzate and Corpas, *Voces diversas*, 89–109.

Vallejo Mejía, Maryluz. "Soledad Acosta de Samper y el periodismo como cátedra de moral." In Alzate and Corpas, *Voces diversas*, 239–59.

Warren, Kim. "Separate Spheres: Analytical Persistence in United States Women's History." *History Compass* 5, no. 1 (2007): 262–77.

2

Railroads and Steamships

Foreign Investment in the Early Development of Peruvian Tourism, 1900–1930

FERNANDO ARMAS ASÍN

As a market economy and industrialization gained ground during the nineteenth century, innovation in transport and communications technology afforded new opportunities for travel. Steamships and railways led the way in connecting continents and various regions of the world. However, the relatively high cost of visiting places outside of Europe and North America ensured that tourism largely remained an elite endeavor. This was a situation that soon began to change as expanding commercial activity coupled with growing interest in foreign lands and cultures required that international travel be made more accessible. In considering the history of tourism in South America around the turn of the twentieth century, this chapter focuses on the role that British and U.S. firms played in developing travel industry infrastructure in southern Peru.

Peruvian Interests, Southern Regional Interests

Responding to promotional literature produced by a number of different publicity outlets and organizations, travelers from the United States and Europe started to make their way to the city of Cusco and neighboring southern Andean destinations early in the twentieth century. Established in 1890, the International Union of American Republics (Unión Internacional de Repúblicas Americanas), for example, distributed a variety of travel guides, such as Albert Hale's 1909 *Practical Guide to Latin-America, Preparation, Cost, Routes and Sight-Seeing*. Other publications included Annie S. Peck's 1913 work, *South American Tour*, and 1910's *A Search for the Apex of America*, which provided a captivating narrative describing her 1908 ascent of Mount Huascarán (in the Cordillera Blanca, approximately 300 kilometers or 185 miles north of Lima). Various celebrities also stimulated international interest in Peru by publishing their travel experiences in an assortment of North American and European outlets.[1]

At the same time, Peruvian intellectuals and politicians stressed the importance of tourism in the development of the southern regional economy. A 1905 study authored by Hildebrando Fuentes, for example, predicted that the city of Cusco—even with Machu Picchu not yet incorporated into the national patrimony—would become the "Mecca of America." With considerable foresight, the lawyer from Lima envisioned tourism as a viable path to regional development in a larger national economy he saw as otherwise hindered by a lack of agricultural modernization, a generally stagnant domestic market, and an apparent lack of natural resources. Fuentes imagined tourism to Peru would bring people and tradition together to create "the true Rome, the Eternal City where visitors eager for impressions and knowledge, wise archaeologists, heraldic fans and simple tourists would come."[2] As evident in Fuentes's writings, discussion among Peruvian elites on how best to develop the country was marked by a desire for modernization as well as an abiding concern for national patrimony. In a debate that would to some degree engender an exaltation of the pre-Hispanic past and thereby anticipate the coming indigenist movement, not only would intellectuals and politicians from Lima participate but so also would a number of regional elites from Cusco, Puno, and Arequipa.[3]

In making their way to the Southern Andes for business and pleasure travel, foreign tourists typically entered Peru via two main routes: either through the port of Callao and the city of Lima or by way of the southern port city of Mollendo. U.S. maritime travel generally originated in either New York or some other East Coast U.S. city and proceeded to the Panamanian port of Colón for transfer by rail to the Pacific coast (before the opening of the Panama Canal in 1914). Those coming from Europe journeyed by way of the South Atlantic, making their way through the Straits of Magellan to the Chilean port of Valparaíso and then up the coastline to either Mollendo or Callao. A variety of shipping companies, including the U.S.-based Grace Line, the English Pacific Steam Navigation Company, and the Chilean South American Steamship Company (Compañía Sudamericana de Vapores), among others, provided regular service for international travelers.

In those years it was no longer just foreign individuals or family groups who visited those sites. In January 1909 Cullver Tour Company of Boston organized and then brought a group of North American businessmen and bureaucrats to Peru, with their families in tow. Headed by Samuel G. Cornell on a tour that would eventually span the globe, the entourage departed from New York City for Panama and then sailed to Lima aboard the steamship *Huasco*. After a stay at the Hotel Maury in the city, the group ventured south, following a route that included Mollendo, Arequipa, and Puno, before heading to La Paz, Buenos Aires, and then on to England before returning to New York.[4]

It became increasingly common for various groups of elite tourists to arrive by boat and head to Lima and then tour the Pacific coast area before continuing on to Mollendo. The presence of two such tour groups prompted an enterprising photographer to create a souvenir album titled *West Coast Pacific*.[5] In general, a number of travel agencies, in cooperation with different shipping companies, became very interested in promoting travel to Peru as part of a South American tour that included stops in Panama, Ecuador, Bolivia, and Chile.

Even the shipping companies began a program of tourism promotion. For example, an advertisement for the Pacific Steam Navigation Company in 1911, apart from offering its regular commercial services, also mentioned a forthcoming "tour round South America," while the Grace Line highlighted amenities available to trav-

6. Neptune party, Grace Line steamer, 1923. Wikimedia Commons.

elers on their ships, including pleasant cabins and a ready medical staff of "nurses."⁶

To the Southern Andes and beyond by Steamship and Railroad

With construction of the Southern Railroad in 1871, travelers making their way to the Andes would typically arrive by sea at Mollendo and stay at either the Hotel 4 de Julio or a local *posada*. Starting in 1890, visitors would journey via the British-financed Peruvian Corporation railroad to Arequipa and then on to Puno and then, after 1893, to Sicuani on their way toward Cusco. Trains were still the only modern transport available.

As the Southern Railroad was not turning a profit in transporting either Peruvian merchandise or passengers, it was thought in the beginning of 1890 that the Bolivian market should be tapped.⁷ The Desaguadero River was dredged to facilitate commercial traffic across Lake Titicaca from Puno to Oruro and Corocoro by the commercial vessels *Yapura* and *Yavari*. Before long the steamer *Coya* and later the *Inca* assumed these same routes.⁸ Cargo and passengers traveling from La Paz by train could head toward the Bolivian port of Guaqui (at the southwest corner of Lake Titicaca) and then be transported by ship to Puno and then on to Mollendo. Small boats that served villages and stopped at docks along the Titicaca lakefront provided access for the shipment of local goods and passengers—a situation that would soon prove amenable for the incipient tourist trade.

In the interest of adding services to increase revenue, the Peruvian Corporation—much like other railroad companies—saw tourism as a potential source of profit. The opening of the Panama Canal in 1914 helped encourage the company to promote its two main train routes as a way for excursionists to travel to any number of attractive sites. Publicity was directed toward the North American public as well as a diverse range of foreigners traveling the South Pacific. In the case of the southern line, it was said that travelers simply could not proceed "without paying a visit to the beautiful Arequipa" (aka the White City) to see the beauty of the Misti volcano off in the distance while also paying a visit to some of the city's attractive churches and scenic avenues. Indeed, Arequipa at the beginning of the twentieth century was a city undergoing significant change as part of a larger, expanding Southern Andean regional economy. Major construction dedicated to expanding central plazas and streets was taking place along with a wave of new housing construction that sought to emulate the latest European architectural styles and thereby significantly transform the appearance of the city. A number of modern hotels were built to compete with older inns and established guesthouses. Among the newer facilities were the Grand Hotel Central and the Hotel Europa.[9]

In an effort to attract tourists, the Peruvian Corporation invited the director of the Lima-based English-language publication *Peru To-Day* to write a history (*una crónica*) of Arequipa, along with recommendations to stay in the Hotel Central or to eat at the Arequipa Club. Upon visiting the city, the railroad director commented on a number of new attractions, including the Harvard Observatory.[10] Other well-known sites, such as the Gran Hotel Véliz and the Yura hot springs, added to the increased volume of travelers coming to Arequipa. For 1912 the Peruvian Corporation announced the publication of an advertising supplement, *The Land of the Incas*, intended

7. Map of Peru Southern Railways routes. Courtesy Fernando Armas Asín.

to further enhance the area's appeal to tourists. The promotion not only described the city of Arequipa and its surroundings but also encouraged people to visit other destinations, including Cusco, the "heart of the ancient Inca civilization," Puno, and Lake Titicaca, known as "the highest body of navigable water in the world."[11]

In fact, those who continued their journey on to Puno were able to reach their destination by utilizing a recently established weekly rail service linking the two cities and featuring first-class Pullman passenger cars. Along the route travelers witnessed a host of impressive sites as they journeyed high into the sierra of the Puno region and then enjoyed a visit to Lake Titicaca aboard one of the local steamers.

On a trip to the region a French visitor, Marcel Monnier, wrote about his southern Peruvian rail journey and travel to Lake Titicaca. In making his way across the lake at night aboard one of the steamers, he noted his sense of ecstasy as he looked out on the Bolivian Andes as they went from Sorata Peak to Illamani. Monnier subsequently observed, "There are few panoramic views comparable to this [and] I cannot recall having witnessed something so powerful when traveling—maybe only once—and that was when I was in the Himalayas climbing Kichinjinga in the highlands of Darjeeling." With this, the French traveler further noted, "[How] spectacular it [was] to contemplate the sunset in this part of the world and to see the South Star appear at night."[12]

As the century progressed, tourism to Cusco steadily increased. Setting out from Lake Titicaca, travelers made their way by rail in what today is described as the most beautiful train ride in all of South America—if not the world. Curiously, before 1908, when the rail connection was completed, the route from the lake to Cusco extended only to Sicuani, where sojourners—following a stay at the Lafayette or the Peru Hotel—continued on to the Imperial City by horseback.

In Cusco the city's iconic pre-Hispanic architecture, colonial churches, and administrative structures mix with a colorful assortment of popular *chicha* bars, snack huts, and local inns to create a memorable travel impression. In the early twentieth century, as the tourism industry was just getting started, many local accommodations (*tambos*) stood as "somewhat precarious lodgings that provided hospitality primarily for travelers from close provincial areas" rather than for more well-to-do foreign tourists.[13] Yet with the slow but gradual increase in businesspeople and tourists visiting the city, a handful of commercial hotels were built.

Around 1900 the Europa Hotel, for example, had already begun accommodating guests in the home of Benigno de la Torre, located in the Portal de Panes just off the centrally located Plaza de Haukaypata. By 1903 the Hotel Comercio was open in the old city square with an Italian owner named Poleti. Another facility, known as the London and England Hotel, also provided local hospitality. By 1910 the Hotel Royal, owned by Kalafatovich and Company, had established two sites: one located in the Portal del Espinar a block south of the main plaza and another nearby on Marqués Street.[14]

Back then, traveling to the neighboring Sacred Valley was a real challenge. It was said that to first get to Urubamba, one had to ride for three or four hours on mule- or horseback. From there it similarly took two and a half hours to reach Quispicanchis and then the same amount of time again to Chincheros. Even after 1911, when Machu Picchu was first opened to visitors, it took three days of traveling to make it there, given that the roads were bad—often consisting of only stone-filled trails with a smattering of rustic inns along the way.[15] As in past years, a letter of recommendation or a talk with the parish priest or a local landowner was the best way to obtain a guide and journey to a variety of interesting sites.

Traveling much of the aforementioned territory in 1909 upon his return from the Pan-American Scientific Congress held in Santiago, Chile, Yale University professor Hiram Bingham disembarked in Mollendo and made his way to Arequipa, where he stayed in the luxurious Hotel Moreno, which even then featured indoor plumbing, including hot and cold water in guest rooms. From Arequipa, Bingham went on to Cusco by train. He stayed at a place called Hotel Comercio, where he found the food to be appetizing and the rooms clean. Hotel hospitality aside, Bingham cautioned those who might be making the four-week journey from New York that

8. Hiram Bingham III, *Machu Picchu in 1912*. Wikimedia Commons.

travel could be challenging at times. Sojourners would have to put up with a host of inconveniences, including dirt, noise, kerosene lamps, and a lack of wheeled vehicles, along with a host of other, often unanticipated unpleasantries. Despite this, Bingham encouraged people to make the trek, noting that the availability of Indian porters and the chance to visit marvelous sites such as of Koricancha, Sacsayhuamán, and the city's Inca palaces would prove well worth the effort.[16]

Similar developments occurred in central Peru. The magazine *Peru To-Day* published a regular feature, "Guide for Shoppers and Tourists," that provided travelers with helpful information about staying in Lima and surrounding areas. Advertising for the Grand Hotel in Chosica—an inviting province just northeast of Lima known for its relaxing and restorative atmosphere—helped make it a popular destination for city residents once easy access by train was established. Before long, weekend groups heading in the direction of Chosica also began making their way farther inland. In 1909 a group of university students organized what would be a highly

9. Lima, Oroya, and Huancayo Railroad. Wikimedia Commons.

publicized train excursion into the Central Andes, with stopovers along the way (Ticlio, La Oroya, Huancayo).[17]

The Peruvian Corporation increasingly took notice of the tourist potential of this route and began promoting its own line, the Central Railroad, as the perfect way to visit not only Chosica but also places in the Central Andes, such as Ticlio, Yauli, Tarma, and La Merced. The company soon also promoted the Mantaro Valley and the town of Jauja (known for its dry climate, favored by people suffering from tuberculosis). With Chosica being the easiest place to reach from Lima, the company's initial efforts included regular Saturday service, which began in 1911. Building on this, the Peruvian Corporation advertised rail access to other destinations, such as Cajamarquilla, Santa Eulalia, San Bartolomé, and Huancayo, in *Peru To-Day*.[18]

A growing assortment of regional travel guides soon became available, such as E. Muecke's 1911 publication *The Center of Peru: Scenes along the Central Railway*. This and similar publications provided practical information for tourists along with information regarding purchase of archaeological artifacts and more commercial access to raw materials. As rail infrastructure took hold, not only were places within close reach of the Central Railway incorporated into regional development but so were more remote places, such as Chanchamayo Province, which included the English company town of Perené. As seen elsewhere, tourist guides accompanied other types of published material aimed at drawing people to the area.

The Golden Twenties and the Growth of the Tourism Industry

Thanks to efforts on the part of the Peruvian Corporation and others, the tourism industry in Peru by the early 1920s had taken shape. By the middle of the decade a growing clientele of middle-class, student, and even working-class travelers from Lima had begun making their way to a variety of destinations. Reduced fares (20 percent discounts) to destinations in the central sierra during national holidays proved especially popular.[19] To the south, travel activity received a significant boost from coverage of Hiram Bingham's expeditions to Machu Picchu in *National Geographic* and other publications that commanded international attention.

In the meantime the Southern Railway was on its way to connecting Cusco and Machu Picchu. Service reached Ollantaytambo and Cedrobamba in 1928 and then extending to La Máquina (Machupiccu pueblo or Aguas Calientes). From there it was still necessary to travel up the steep incline by horseback or on foot. Eventually a more substantial roadway passage would provide more convenient access for the growing number of visitors to the site.

Peruvian local elites increasingly took an interest in the tourism industry. Reimagining and packaging the Peruvian past became an important undertaking not just for intellectuals writing books and articles but also for those wanting to engage the emerging travel market. Since the early 1900s, photographers in Cusco had been producing scenes of urban life, monuments, and images of "typical" Indians for purchase. By the 1920s the increasing visitor demand for such photographs and postcards as souvenirs had generated a profitable business. Publication of the *Guía General del Sur del Perú*, containing eighty photographs on the centennial of Peruvian independence in 1921, proved hugely successful. Other illustrated collections soon joined the *Guía General del Sur del Perú*, which was sold in the Cusco bookstore of photographer Héctor Rozas along with an assortment of individual postcards and photos of the city by other local photographers, such as Miguel Chani, Manuel Figueroa Aznar, J. Chambi, and José Gabriel González. Adding to the growing availability of these publications, Albert A. Giesecke, who was a specialist in indigenous culture and rector of the University of Cusco, issued the *Guía de Cuzco: La Meca de América del Sur* in 1924. Giesecke's volume offered twenty-six pages of text and a series of photographs that included a host of panoramic shots and illustrations of nearly every church in Cusco.

Travel publicity disseminated from Lima soon came from another important organization founded in 1924: the Peruvian Touring Club, whose magazine *Perú* was distributed in consulates, hotels, and other key international institutions, thanks to the efforts of the association members. With assistance from staff at the Peruvian consulate in New York City, one article published in *Perú*, titled "Tourism Should Be a Primary National Industry," called upon the club to advertise Peru as a premier tourism destination while at the same time urging government officials to find solutions to whatever administrative or environmental challenges the nation might face in developing the industry.[20]

The following year Peruvian Company promotions began using the slogan "Getting to know the unlimited possibilities of Southern Peru is to add to the greatness of the Republic overall."[21] The campaign made sense because rail passenger traffic in the Central and Southern Andes had given rise to a network of hotels—"railway" hotels that provided accommodation for travelers in Mollendo, Puno, Sicuani, and Cusco in the south and Chosica, Huancayo, Cerro de Pasco, and others in the central region. Judging from various testimonies available at the time, these hotels were all relatively comfortable places, capable of hosting both domestic and international travelers.

The Grace Line also upgraded its services while intensifying publicity efforts. Beginning in 1925, the company offered four South

American coastal tours. One tour involved traveling on the Peruvian Railway running from Mollendo to Arequipa, Juliaca, and then Cusco over a fifteen-day span, with overnight accommodations at a number of private hotels.[22] Meanwhile, Wagons-Lits, said to be the "premier global travel agency," combined forces with the well-established Thomas Cook Agency and soon opened an office in Lima and began organizing tours.[23] By the end of 1928 industry leaders felt that "all indicators suggested that 1929 would be a great year for tourism in Peru."[24]

Depression-Era Adjustments and Development Strategies

The economic crisis that began in 1929 caused a substantial drop in the number of tourists visiting and traveling in Peru. As a result, steamship lines, railroad companies, and travel agencies were forced to develop new strategies. In this highly uncertain economic climate Peruvians adapted to changing market conditions by joining with other international tourism providers.

The Grace Line diversified its business by commissioning cruise ships for service in Central and South America. This shift was complemented by an incentive program specifically aimed at boosting tourism along the Pacific coast. To do this, the company distributed photos and brochures from its headquarters in New York City. It hired journalists to publicize Peru in daily and weekly publications while also contacting a number of individuals well connected in political and tourism circles, including members of the Peruvian Touring Club, who, after opening an office at the Pan American Union in Washington DC in 1933, began screening the film *Cuzco, el Imperio de los Incas* in a number of cities throughout the United States and Europe.[25]

Other international travel companies followed suit. Early in 1932 Bence Tourist Company owner Susana Bence arrived in Lima and quickly made known her desire to become "a great advocate for Peru." She interviewed the minister of foreign relations and other authorities as a way to express her desire to do business. Around the same time, Harold Harris, head of fledgling Panagra Air, began promoting the idea that tourism could make Peru the "California of South America," and he envisioned a number of dynamic economic linkages, including silverwork and other craft exports.[26]

Many challenges remained, however. Darío Eguren de Larrea, author of the internationally popular 1929 work *El Cusco: En su espíritu y sus maravillas* recognized the need for Peruvian state involvement in the burgeoning industry ("something other countries of America already do") and in 1932 called for federal support.[27] A short while later the Argentine director of the Exprinter Travel Service visited Lima. As someone who organized package tours by rail between Argentina and Chile, he wanted to explore the possibility of facilitating a similar kind of product to offer Peruvians and those who regularly traveled in Peru. As no doubt those who worked for Wagons-Lits Cook, the Grace Line, and other international agencies now offering "all-inclusive" trips to Peru were well aware, the Argentine businessman identified several major challenges to industry development. The most significant of these included the fact that the port of Mollendo was in poor condition and could not accommodate cruise ships with three hundred to four hundred passengers each, that no serviceable road had been built from the capital to Cusco, and that few high-end hotels operated in the country outside of Lima.

Despite these and other shortcomings, travel publicity disseminated widely in North America, the Panama Canal Zone, and the Southern Cone during the early 1930s ambitiously promoted a list of places to visit, available lodgings, and existing routes to travel. Anticipating the four-hundred-year anniversary of the founding of Lima, tourism industry insiders in 1935 knew all too well the challenges they faced.

Major infrastructural advancements were necessary if larger-scale international travel to Peru was to become a reality.[28] As was the case in other American nations at the time, Peru needed significant state policy, planning, and federal funding for tourism development.[29] To this end, members of the Peruvian Touring Club, Lima elites, and various regional intellectual groups stepped up their efforts to encourage the federal government to develop hospitality infrastructure, remove remaining legal obstacles to foreign travel, and help advance promotional efforts. In the years to come, these and other important industry concerns would need to be addressed not solely by private interests but with the active engagement of the Peruvian state.[30]

Conclusion

Before the 1930s many Latin American governments played a role in developing tourism in conjunction with elite and middle-class groups, who also helped stimulate national tourism, Argentina being one example.[31] Industry growth in Peru, however, originated primarily with foreign agents. Serving as the essential drivers for international travel to the Central and Southern Andes, incipient tourism strategies on the part of a few key firms were first designed to create travel-related initiatives that would complement larger existing commercial activities. In order to help establish a market for tourism, these foreign entities soon joined forces with a host of Andean regional elites wishing to develop hospitality infrastructure and services in a variety of local areas. This was a kind of cooperative international process, undertaken during the early decades of the twentieth century, and it was a pattern replayed in other countries along the Pacific coast, such as Chile, as well as in an assortment of Caribbean nations.[32]

In the case of Peru a more robust national tourism market less dependent on foreign capital would eventually take shape but not until after 1930. Still centered mostly in Lima, activities on the part of a new generation of domestic tourism and travel service providers gradually coalesced to act as a counterweight to industry activity previously dominated by foreign companies.[33]

One might ask why it was that tourism developed in Peru in a manner initially more dependent on external capital flows and market linkages. From a historical point of view, industry growth took shape the way it did because of the absence of a sizable and dynamic middle class. Had such a group existed, it would have been able to shape and sustain an internal market and thus a more nationally directed process, with foreign corporations acting largely at the behest of Peruvian state and commercial leadership.

Notes

This chapter was translated from Spanish by Andrew Grant Wood with assistance from Bruce Dean Willis.

1. Armas Asín, *Una historia del turismo en el Perú*, 1:146.
2. Fuentes, *El Cuzco y sus ruinas*, 238.
3. Armas Asín, *La invención del patrimonio católico*, 85, 93.
4. *Peru To-Day* 1, no. 2 (1909): 7.
5. Juan Carlos La Serna, interview by author, Lima, Peru, December 20, 2015.
6. Ads in *Peru To-Day*, vol. 3 (1911); ads in *Touring Club Peruano*, vols. 1–4 (1925–30).
7. Bonilla, *Gran Bretaña y el Perú*, 4:241.
8. Bonilla, *Gran Bretaña y el Perú*, 4:29; *Variedades*, no. 73 (August 24, 1909): 500.
9. Bonilla, *Gran Bretaña y el Perú*, 4:3, 86; García Euribe, *Aproximación al estudio de la evolución*, 82; *Peru To-Day* 3, no. 9 (1911).
10. *Peru To-Day* 3, no. 9, (1911): 19, 23–25.
11. *Peru To-Day* 3, no. 5 (1911).
12. Monnier, *De los Andes hasta Pará*, 118–21. E. W. Middendorf similarly provides a description of traveling by boat by way of Titicaca lakeside villages. Middendorf, *Perú*, 3:237–334.
13. Tamayo Herrera, *Historia social del Cuzco*, 298.
14. García Euribe, *Aproximación al estudio de la evolución*, 82; Tamayo Herrera, *Historia social del Cuzco*, 298–99; Valcárcel, *Memorias*, 22–43.
15. García Euribe, *Aproximación al estudio de la evolución*, 82.
16. López Lenci, *El Cusco, paqarina moderna*, 38–42; Porras Barrenechea, *Antología del Cusco*, 331–44.
17. *Variedades*, no. 81 (October 23, 1909): 807–8.
18. *Peru To-Day* 3, no. 3 (1911): 35.
19. *Touring Club Peruano*, vols. 1–4, 1925–30.
20. Armas Asín, *Una historia del turismo en el Perú*, 1:198.
21. *Touring Club Peruano* 1, no. 1 (1925). *Touring Club Peruano* was the official publication of the club.
22. *Touring Club Peruano* 1, no. 1 (1925).
23. *Touring Club Peruano* 1, no. 1 (1925); vol. 3 (1927). Created in 1927 and lasting well into the 1970s, Wagons-Lits Cook dominated the European market, just as American Express commanded much of the North American travel business. In 1928 the wholesale Exprinter Travel Service was also established, as was the Expreso Villalonga. Khatchikian, *Historia del turismo*, 228.
24. *Memoria del Segundo Congreso Sudamericano de Turismo*, 2:270. See also *Touring Club Peruano* 4, no. 44 (1928): 19–24; 4, no. 45 (1928): 5–6, 19; 5, no. 55 (1929): 43; 6, no. 61 (1930): 47; and 6, no. 62 (1930): 33–36.
25. *Touring Club Peruano* 6, no. 72 (1931): 25; 8, no. 82 (1933): 5.
26. *Touring Club Peruano* 6, no. 72 (1931): 25.
27. *Touring Club Peruano* 7, no. 76 (1932): 10.

28. *Touring Club Peruano* 8, no. 80 (1933): 27; 8, no. 82 (1933): 14; 9, no. 83 (1934): 13; 9, no. 84 (1934): 7.

29. *Turismo*, no. 112 (1937): 7.

30. Armas Asín, *Una historia del turismo en el Perú*, 1:219–30; *Touring Club Peruano* 7, no. 73 (1932): 14.

31. Pastoriza, *La conquista de las vacaciones*, 249–58; Piglia, *Autos, rutas y turismo*, 225–36; Valenzuela Valdiviezo and Coll-Hurtado, "La construcción y evolución."

32. Booth, "El Estado Ausente," 107–23; Skwiot, *Purpose of Paradise*, 206–14; Ward, *Packaged Vacations*, 200.

33. Armas Asín, *Una historia del turismo en el Perú*, 1:137–64; Armas Asín, "Autos, caminos, y clases medias."

Bibliography

Armas Asín, Fernando. "Autos, caminos, y clases medias en los años veinte: Entre el ícono cuzqueño y el desarrollo del turismo nacional." *Turismo y Patrimonio* 11 (2017): 113–37.

———. *La invención del patrimonio católico: Modernidad e identidad en el espacio religioso peruano (1820–1950)*. Lima: Asamblea Nacional de Rectores, 2006.

———. *Una historia del turismo en el Perú: El Estado, los visitantes y los empresarios (1800–2000)*. 2 vols. Lima: Universidad San Martín de Porres, 2018.

Bonilla, Heraclio, comp. *Gran Bretaña y el Perú: Informe de los Cónsules Británicos; 1826–1900*. 5 vols. Lima: Instituto de Estudios Peruano y Banco Industrial del Perú, 1975–77.

Booth, Rodrigo. "El Estado Ausente: La paradójica configuración balnearia del Gran Valparaíso (1850–1925)." *Eure: Revista Latinoamericana de Estudios Urbanos Regionales* 27, no. 83 (May 2002): 107–23.

Boyer, Marc. *L'invention du tourisme*. Paris: Gallimard-Découvertes, 1996.

Fuentes, Hildebrando. *El Cuzco y sus ruinas: Tahuantinsuyoc kapacllacta*. Lima: Imprenta La Industria, 1905.

García Euribe, Rafael. *Aproximación al estudio de la evolución del espacio de hospedaje en el turismo peruano*. Lima: Universidad de San Martín de Porres, 2015.

Giuntini, Andrea. "Ferrocarriles y turismo en Italia desde los inicios del ochocientos hasta la introducción de los 'trenes populares' en la época fascista." *Historia Contemporánea*, no. 25 (2002): 101–23.

Khatchikian, Miguel. *Historia del turismo*. Lima: Universidad de San Martin de Porres, 2000.

Löfgren, Orvar. *On Holiday: A History of Vacationing*. Berkeley: University of California Press, 1999.

López Lenci, Yazmín. *El Cusco, paqarina moderna: Cartografía de una modernidad e identidades en los Andes peruanos (1900–1935)*. Lima: Universidad Nacional Mayor de San Marcos y Consejo Nacional de Ciencia y Tecnología (Concytec), 2004.

Memoria del Segundo Congreso Sudamericano de Turismo: Antecedentes. 2 vols. Lima: Empresa Editorial "Cervantes," 1929.

Middendorf, Ernest W. *Perú: Observaciones y estudios del País y sus habitantes durante una permanencia de 25 años.* 3 vols. Lima: Universidad Nacional Mayor de San Marcos, 1974.

Monnier, Marcel. *De los Andes hasta Pará, Ecuador-Perú-Amazonas.* Lima: Instituto Francés de Estudios Andinos y Banco Central de Reserva del Perú, 2005.

Morgan, Nigel J., and Annette Pritchard. *Power and Politics at the Seaside: The Development of Devon's Resorts in the Twentieth Century.* Exeter, UK: University of Exeter Press, 1999.

Pastoriza, Elisa. *La conquista de las vacaciones: Breve historia del turismo en la Argentina.* Buenos Aires: Edhasa 2011.

Peru To-Day, 1909–11. English-language monthly magazine published in Lima, Peru, and edited by John Vavasour Noel.

Piglia, Melina. *Autos, rutas y turismo: El Automóvil Club Argentino y el Estado.* Buenos Aires: Siglo XXI, 2014.

Porras Barrenechea, Raúl. *Antología del Cuzco.* Lima: Librería Internacional del Perú, 1961.

Skwiot, Christine. *The Purpose of Paradise: U.S. Tourism and Empire in Cuba and Hawai'i.* Philadelphia: University of Pennsylvania Press, 2010.

Tamayo Herrera, José. *El pensamiento indigenista.* Lima: Mosca Azul, 1981.

———. *Historia del indigenismo cuzqueño, siglos XVI–XX.* Lima: Instituto Nacional de Cultura, 1980.

———. *Historia social del Cuzco republicano.* Lima: Industrial Gráfica S.A., 1978.

———. *Historia social e indigenismo en el Altiplano.* Lima: Ediciones Treintaitrés, 1982.

Tissot, Laurent, ed. *Construction d'une industrie touristique aux XIXe et XXe siècles.* Neuchâtel, Switzerland: Alphil, 2003.

Touring Club Peruano: Revista del Touring Club Peruano. Lima, Peru, 1925–34.

Turismo. No. 112. Lima, Peru, 1937.

Valcárcel, Luis E. *Memorias.* Cusco: Ministerio de Cultura, Dirección Desconcentrada de Cultura de Cusco, 2015.

Valenzuela Valdiviezo, Ernesto, and Atlántida Coll-Hurtado. "La construcción y evolución del espacio turístico de Acapulco." *Anales de Geografía de la Universidad Complutense* 30, no. 1 (2010): 163–90.

Variedades: Revista Ilustrada. Lima, Peru, 1909.

Walton, John K. *The British Seaside: Holidays and Resorts in the Twentieth Century.* Manchester, UK: Manchester University Press, 2000.

———. *The English Seaside Resort: A Social History, 1750–1914.* Leicester, UK: Leicester University Press, 1983.

Ward, Evan R. *Packaged Vacations: Tourism Development in the Spanish Caribbean.* Gainesville: University Press of Florida, 2008.

3

Changing Caribbean Routes

The Rise of International Air Travel

BLAKE C. SCOTT

How many times have you stood on the deck of a steamer, tossing in a rough sea and enviously watched the gulls wheeling and dipping 'round the vessel. What swiftness and lightness, what ease, while you suffered the agonies of the endless rolling and pitching of a spiteful sea. How you longed for the smooth, quick flight of the gull.... But now man has mastered the principles of flight and may enjoy the comfort, speed and safety of aerial transportation.

—*The Air-Way to Havana*, Pan American World Airways brochure

That is the terrible thing—the curious illusion of superiority bred by height.

—ANNE MORROW LINDBERGH

On January 16, 1928, Pan American World Airways launched the first international passenger flight from the United States—flying ninety miles from Key West, Florida, to Havana, Cuba. Within three years Pan Am had established a nearly twelve-thousand-mile system linking twenty-three countries in the Americas. By the early 1940s the airline's reach was global: flying to sixty countries and covering an impressive ninety-eight thousand route miles. As Pan Am executives later recalled, "our first flight [to Havana] was a short one . . . but it was long on history, as it opened America to worldwide air travel."[1]

10. Pan American World Airways first passenger flight, January 16, 1928. University of Miami Special Collections.

The rise of international flight is a well-documented history, yet most authors tend to tell narratives of "technological triumph" and focus less on the complex cultural impacts when considering the development of air transportation.[2] For most modern travelers, flight has become a taken-for-granted aspect of mobility. In doing so, however, they often ignore the fact that in less than a one-hundred-year period, the act of overseas travel significantly morphed from precariously sailing at sea to flying comfortably thousands of feet above the surface of the earth. The revolution in transportation that gave rise to modern tourism embodies not only technological change but a much wider transformation in how travelers experience time, space, and cultural connection. Aviation, in other words, created a powerful force for "the annihilation of space by time" and, as a result, dramatically altered international tourism.[3]

Travels across the Gulf Stream

The transition from maritime to air travel was a huge leap in hemispheric relations and, more broadly, in the history of globalization. Flight radically altered the way people crossed borders, but it did so in ways today not well remembered. The phenomenology of travel has changed over time: the speed, distance, time, relation to natural elements, smells, sights, and sounds are significantly different today when compared to travel experiences a century ago. Modern aviation quickened the journey, but it also contributed to a growing disconnect between people and the environment.[4]

Technological advances fostered the sense that the ninety miles separating the United States and Cuba amounted to merely a short "hop." This perspective, however, offers a new way of imagining what was long considered a potentially perilous journey. The shift from sea to air mobility had profound implications for how people experienced and imagined travel to the Caribbean. Consider, for example, one well-known and dramatic journey from Florida to Cuba just thirty years before Pan American Airways began flying that route.

On New Year's Eve 1896 the writer Stephen Crane boarded the ss *Commodore* in Jacksonville, Florida, headed for Cuba. Crane was on his way to cover the outbreak of war on the island. But in rough seas off the central Florida coast, the *Commodore* began to take on water. With the ship slowly sinking, the surviving crew made it into a small boat, which they struggled to keep afloat. Crane later wrote of his ordeal in a short story called "The Open Boat":

> None of them [including himself] knew the color of the sky.
> Their eyes glanced level, and were fastened upon the waves that swept toward them. These waves were of the hue of slate, save for the tops, which were of foaming white, and all of the men knew the colors of the sea. The horizon narrowed and widened, and dipped and rose, and at all times its edge was jagged with waves that seemed thrust up in points like rocks. . . . These waves were most wrongfully and barbarously abrupt and tall, and each froth-top was a problem in small boat navigation.[5]

Although most sea journeys in the late nineteenth century did not end in disaster like Crane's, his story nevertheless highlights the vulnerability and excitement of "going to sea." The possibility of being lost in a storm, blown off course, or shipwrecked was a concern deeply embedded in the cultural experience of maritime travel. Indeed, the waters off Florida and Cuba were historically a treacherous sea for sailors and fishing crews. "The Gulf Stream," as another travel writer liked to tell readers, was "the last wild country there is left."[6]

Travelers took to the ocean dreaming of distant lands, but the journey itself "represented an important 'site of imagination.'" It was, as the historian Eric Zuelow describes, "a third place that existed between one's point of origin and one's terminus."[7] When sailing from the north to the torrid zone (the tropics), one could observe the first signs of environmental difference not after touching land but in seeing the ocean's condition change. The water became a deep blue color and the wind warmed. If the weather was fair, unlike in Crane's experience, it was a moment of celebration. Seasoned sailors who entered the tropics safely would often practice rituals of "tropical baptism," such as bathing new crew with warm seawater and supplying them with strong drink. Such rituals when entering the equatorial zone highlight how "the qualities of its waters, its weather, its place in relation to the sun and stars, even its sea life," were fundamental to the way Europeans and Euro-Americans understood travel to the tropics.[8] Long-distance mobility—from the earliest colonial encounters to the beginning of the twentieth century—depended on the nature of the ocean's power.

Going to sea was a risky and therefore revered affair. According to UNESCO, some three million shipwrecks lie on the ocean floor; the seabed in the Straits of Florida, one of the most heavily traveled shipping lanes in the world, is particularly riddled with sunken vessels.[9] During much of maritime history it was difficult to predict approaching weather, precisely identify one's location, or even know the depth of water beneath one's vessel. An impediment to international travel, especially in leisure, is of course the ability to get to a destination. As a result, mariners invented an astounding and creative multitude of navigational technologies, such as the sextant (circa 1731), which measures the angle between an astronomical object and the horizon (celestial navigation) and thus the distance between two points on earth.[10] Competing imperial powers—Spanish, Portuguese, Dutch, English, French—had for centuries funded maritime expeditions, research, and design to support political, military, and commercial expansion in the Americas. The geopolitical interests of empire relied on maritime travel. From Amerigo Vespucci and the mapping of the Americas to Charles Darwin and the South American Pacific coast, science and technology at sea were intimately entangled with imperial expansion. Efforts to improve maritime transportation, including the later development of coal-powered steamships, were critical in making the modern world.[11]

Pan Am and the Age of Aviation

Air travel contributed to a wide array of social and cultural changes—not all of them particularly exhilarating. Writing about his experience on a Pan Am flight for the *Havana Evening Telegram* in early 1928, journalist A. Edward Stuntz, remarked, "A trip from Havana to Key West and return by one of the big Fokker planes of the Pan-American Airways is novel enough but not exciting." Rather, he claimed boredom, observing, "Now I've heard about squalls in the air and I'm disappointed. . . . Here in the cabin, there is no dramatic business. It is all about as fearful as a trolley ride and not half as bumpy." Stuntz, in fact, apologized to readers hoping for more adventure: "I am sorry I couldn't adequately write this trip from the air. But how can anyone conjure the proper phraseology for a sparklingly interesting experience in an hour and a half. Let's call it short and sweet and let it go at that."[12] Taking the extremes of Crane's near-death experience at sea and Stuntz's aeronautical indifference, one can, while traveling the same route, glimpse how dramatic a shift it was to move from overseas mobility dependent on the ocean to traveling thousands of feet up in the air.

With advances in air travel, the traveler's experience and sense of the sea became—both literally and figuratively—vertically distant. Modern flight is a form of sensory dislocation, or in the words of anthropologist Arjun Appadurai, "deterritorialization."[13] The natural elements, while physically still in place, had little to do with the air traveler's perception, or lack thereof, of the journey. Air travel

11. Cover of *New Horizons* magazine featuring the iconic painting *Yankee Clippers Sail Again*, by Gordon Grant. University of Miami Special Collections.

increasingly isolated passengers from the outdoors. Ask any traveler today, for example, about what they do on an overseas flight and the list of activities is rather standard: sleep, watch a movie, listen to music, drink, eat, read, maybe talk, and rarely getting out of one's seat except to use the lavatory.[14] In contrast, at the level of the sea untold generations of sailors have witnessed the tiniest of ecological events—a flying fish, a visiting bird, a ship in the distance, a change in the wind or waves.

In addition to changing travel experiences of space, flying shortened the time of transit. Depending on the ship and the weather, sailing from Key West to Havana could take anywhere from fifteen to twenty-five hours. Meanwhile, by steamship the trip could take a full day at sea. In the late 1920s and early 1930s, however, air travel reduced trip time by at least fivefold compared to journeys via engine-powered ships and fifteenfold compared to traditional wind-powered vessels. Flight, as "time-space compression," rejected the significance of the ocean's territoriality. "The air," as *Fortune* magazine argued as early as 1943, "is a [new] blue-water ocean to which every nation potentially has access for trade and high strategy in all directions. Under its intoxicating implications the ancient ideas of a world divided by land and sea seem to be as outmoded as the Chinese wall."[15] With the new "logic of the air," travelers were no longer bound by or aware of the geography and environment between their point of origin and their destination. The journey by plane was an abrupt departure from millennia of long-distance travel. This transformation, however, did not fully erase the social practices of the old sea-travel culture.

The Key West–to–Havana air route demonstrates how the combination of technology and geopolitics set the stage for the rise of aviation. Havana was only a short flight from Key West, but the island of Cuba was also a protectorate or, as critics understood it, a neocolonial possession of the United States. Getting people and information quickly from the mainland to the island was useful to imperial governance. Federal contracts to deliver mail paid for Pan Am's early operations, and thus it was no coincidence that the location of U.S. military bases abroad guided the development of

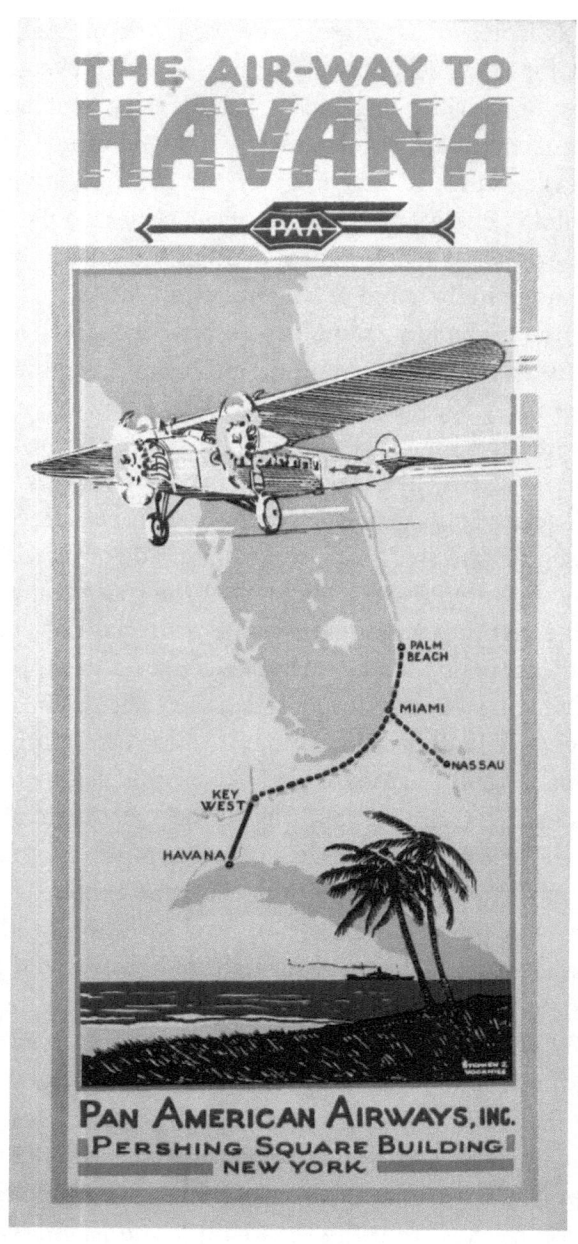

12. Pan American World Airways brochure, *The Air-Way to Havana*, promoting flights from southern Florida to Havana, Key West, and Nassau. University of Miami Special Collections.

Pan Am's routes.[16] This relationship was part of a historical pattern linking state power and transportation advancements.

At the start of World War I in 1914, however, the U.S. military owned a mere 49 aircraft. Strategic interest soon resulted in a wartime aviation boom, initiated in March 1915 when Congress used the Naval Appropriations Act to create the National Advisory Committee on Aeronautics—later to become the National Air and Space Administration (NASA).[17] By November 1918 U.S. firms had manufactured a whopping 13,894 airplanes, and the number of production firms had expanded from 16 (employing some 168 workers) to more than 300 companies (with approximately 175,000 laborers). Airplane design evolved rapidly as well, transforming slow, wood-and-wire-frame fabric-covered biplanes into sleek, aerodynamic aluminum monoplanes capable of landing on water or land. The speed of these new aircraft, moreover, doubled from a maximum velocity of 75 mph to upwards of 150 mph in just a few years.[18]

In 1919, less than three months after the Treaty of Versailles, Germany (despite its defeat) began the world's first passenger airline service, with routes linking Berlin, Leipzig, and Weimar. The British and French also converted former military bombers to run commercial service between London and Paris. There was as well a brief period of European investment in regional air travel in the Americas. Peter Paul von Bauer, a German ace pilot in World War I, opened the first South American commercial airline, the Colombian-German Air Transport Association. In the early 1920s the U.S. aviation industry, domestically and abroad, still lagged behind German, French, and British efforts. However, increasing business investment and military involvement in Latin America and the Caribbean encouraged the U.S. government to develop its own international air transport system. Between World War I and the mid-1920s, U.S. foreign investment grew by nearly 300 percent in Latin America.[19]

In support of U.S. military and economic expansion, Pan American Airways received the first U.S. federal contract to deliver air mail to the Caribbean, beginning with Cuba. The airline was originally founded by four military aviators, veterans of World War I, but after a series of selling and buying, the company was taken over by Juan Trippe (formerly of Colonial Air Transport). Trippe would run the airline for the next four decades, cultivating close government ties. He negotiated to assure Pan Am's position as the U.S. government's "exclusive international" carrier. He claimed proudly that his airline was the "chosen instrument" of the state.[20]

Pan Am carried passengers, mail, and express shipments vital to the U.S. government, but it also worked as a symbol of American power. The airline's first international passenger flight was timed to coincide with President Calvin Coolidge's visit to Havana for the sixth Pan American conference. President Coolidge and Cuban president Gerardo Machado greeted the plane when it landed, claiming it a sign of hemispheric unity. Thousands of U.S. officials, Latin American dignitaries, and tourists were also in Havana that winter season.[21]

The next major Pan Am route, after the Florida-Cuba flights began, was to the Isthmus of Panama. The U.S. government had recently completed the Panama Canal (1914), and hundreds of thousands of sailors and soldiers annually passed through or were temporarily stationed on the isthmus in the 1920s. There was a constant passage of naval and merchant vessels.[22] Pan Am's role was to support the U.S. colonial outpost, assuring fast and efficient communication between canal zone officials in Panama and leaders back in Washington.

In February 1929 aviator Charles Lindbergh—Pan Am's newly appointed technical advisor—began his survey of the route. Departing Miami at 6:00 a.m., on the eve of his twenty-seventh birthday, Lindbergh flew first to Havana and then across the tip of Cuba to the coastline of Belize (then British Honduras). After an overnight stop, he continued on to Honduras, Costa Rica, and Nicaragua before arriving in Panama (covering two thousand miles in nineteen hours of flying time). Colonel Lindbergh's survey flight not only provided expert advice to Pan Am but also generated lucrative publicity. Lindbergh, who had completed the first solo nonstop transatlantic flight two years earlier, was an aviation celebrity. "Whenever Colonel Lindbergh moved," according to reporters covering his journey, "there was an eager crowd following him: persons were eager to shake his hand or even be close enough to have him brush past them."[23]

Pan Am and its pilots were carving out a new American "frontier" of aviation. The airline's regularly scheduled flights in the 1930s and 1940s still depended on the sea, however. In 1934 Pan Am opened Dinner Key Terminal in Biscayne Bay, Miami, the largest marine air facility in the world. The new terminal's fleet of aircraft were called "flying boats," and Pan Am called its planes "Clippers," a nod to the heyday of fast sailing ships that dominated transoceanic trade in the mid- to late nineteenth century. Pan Am's Clippers were amphibious, meaning they could take off from and land on water. They were designed in this way because many of Pan Am's destinations in Latin America and the Caribbean had yet to develop the runways and infrastructure necessary for terrestrial landing. The culture of air travel mimicked not only the technology and geographic routes of maritime travel but also its traditions of expansion and training. Flight crews rigorously trained as if they were part of a ship's crew, learning seaplane anchorage, marine tides and currents, celestial navigation, and over-water navigation.[24] The discipline of sea travel provided the foundation for aeronautics.

Pan Am's marine air terminal in Miami served fifty thousand passengers annually in the 1930s and 1940s, and it attracted even more visitors. By 1935, just eight years after its inaugural flight, Pan Am had also begun to expand beyond the Americas—to Hawaii, the Philippines, and China—and in 1939 it inaugurated regular transatlantic service to England and continental Europe. Each of these developments was closely followed and reported on via radio, newspapers, and newsreels. As Pan Am expanded, so did public interest.[25]

The ports of the past, linking distant places across the ocean, were giving way to the "air" ports of the future. Yet for most travelers in the decades before World War II, travel by ship remained the more affordable and reliable means of venturing abroad. Air travel was expensive, and although the trip was fast, sometimes mundane, and seemingly detached from the natural elements, it was also seen as risky. Stories of aviation catastrophe often captured headlines. In 1927, the year before the inauguration of Pan

Am's passenger service, nearly half of all pilots "attempting long distance flights . . . either lost their lives or had their flights end in near disaster."[26] As one early airline executive admitted, "he never flew in his company's planes when he could avoid it—and his wife flatly refused to fly at all."[27] Americans had gone "plane crazy," but many were still uncertain about boarding one.

World War II and its unleashing of technology brought about safer, faster, and more affordable access to air travel. Following the war, with Europe in ruins, the United States became the world's leader in aviation technology. The Boeing Company, which had developed Pan Am's flying clippers and later the U.S. bombing fleet, came out of the war committed to bringing jet travel to the masses. In May 1954, after years of planning, Boeing opened a massive factory in Renton, Washington, for the development of its 707 jet plane. At the factory's opening eight thousand employees and five hundred community leaders gathered, waving American flags as a brass band played the air force anthem. Pan Am, still the government's exclusive international airline, was the first to use Boeing's new jets for passenger service. In the first year of its jet service in 1958, Pan Am's 707 fleet of twenty aircraft carried more than a million passengers. The 707 had capacity for 150 passengers and could travel at speeds of up to 600 mph (faster than commercial jets today).[28] Passengers and pilots alike commented on the smooth, relatively noiseless journey.

In thirty years (1928 to 1958) Pan Am had gone from one thousand passengers annually to more than a million. International travel had entered the modern jet age, but this change was not the result of technological "manifest destiny." Instead, it was the result of precedent-setting travel routes, U.S. overseas military interests and investments, and postwar consumer affluence.

The expansion of jet travel in the 1960s also led directly to the demise of the era of ocean liners. Maritime passenger service could not compete with the speed and affordability of the jet age. It would be decades before the shipping industry repackaged itself as luxury cruise ships—becoming a "throwback" novelty in comparison to increasingly routine air travel in the late twentieth century.[29]

Flight, like so much of consumer culture, changed—from being an elite, expensive, and novel activity to becoming a mainstay of middle-class culture.

The New Era of Globalization

Scholars of contemporary globalization have argued that the technology of air travel intensified social interactions, bringing once isolated communities into more direct and intimate contact. The airplane has been heralded by both admirers and critics as one of the most important manifestations of the global era. Industry agrees; as U.S. Airways told its customers, "It is a big world. We've got it covered." When reimagined in historical context, the statement leads to the question, Is the world really so "covered?" Compare for a moment the age of sea travel with the age of aviation, as provocatively described by journalist Dea Birkett: "The borders of the world's greatest ocean have been joined as never before. And Boeing has brought these people together. But what about those they fly over, on their islands five miles below? How has the mighty 747 brought them greater communion with those whose shores are washed by the same water? It hasn't, of course. Air travel might enable businessmen to buzz across the ocean, but the concurrent decline in shipping has only increased the isolation of many island communities."[30] This leads to three concluding observations about the effects of "flyover" culture and its impact on international tourism. First, the ability to fly direct from New York City or Miami to Havana, Cuba, may be convenient to the traveler, but it disconnects communities along the route. Two cities gain a direct link allowing for both tourism and emigration, while the entire U.S. Eastern Seaboard, along with communities in the Florida Keys, the Bahamas, and farther south, disappear from the traveler's experience. People, ideas, and goods departing and arriving, from one port to the next, become moored to the past. Technology and wealth have as a consequence become concentrated in fewer and fewer locales. Meanwhile, once bustling port and tourist towns like Port Antonio, Jamaica; Santiago, Cuba; and Colón, Panama, economically decline and lose employment. The rise of air travel for many communities has actually been a process of geographic isolation. Rather than the story of increasing unity, globalizing technology can also lead to disconnection creating vast swaths of "flyover" territory.[31]

Second, for the curious traveler: Is it possible to be more conscious of the world thirty thousand feet above its surface? The energy exerted by traveling across space varies depending on whether one walks, sails, rides, drives, or flies over it. To walk the terrain or, in the case of overseas travel, to sail the sea is, in the words of philosopher Walter Benjamin, "to know the new prospects of his inner being that the text, that lane through the ever-denser internal jungle, opens up."[32] The higher the technology mediating one's experience, Benjamin argues, the more difficult it can be to understand the magnitude or the magnificence of the immanent world. For many travelers this is what made the hardship of the sea worthwhile and made the Gulf Stream the scene of both adventure and literary imaginings. In the case of land-based travel, a similar phenomenological shift also occurred in the nineteenth century, as railroads took over from equestrian mobility, and in the twentieth century, when the automobile outpaced them both.[33]

In the modern era, with less awareness of the environment and the communities between destinations, the Caribbean region in particular has become a sort of sanitized marine park. In the words of Derek Walcott, it is "a blue pool [seen from a distance] into which the [American] republic dangles the extended foot of Florida as inflated rubber islands bob, and drinks with umbrellas float towards her on a raft."[34] The Caribbean, once seen as an adventurous and dangerous sea, has been reinvented as a calm, crystal clear lake of paradisiacal islands welcoming tourists.

Finally, the speeding up of travel may also mean the loss of "real time." If one can get to a Caribbean destination in just a few hours from the mainland United States, there becomes less need to stay. "36 Hours," as the *New York Times* tells readers, is more than enough.[35] As we fly (rush), from work to leisure and back, there is a deep and profound shrinking of time to interact with local communities and ecologies. Aviation technology has been an excellent vehicle of mobility but can also have unanticipated social consequences. The old adage "it's not the destination that matters; it's the journey" has been turned upside down.

Notes

1. Newspaper and Pan Am public relations clippings from "First Passenger Flight, Key West–Havana, January 16, 1928," Folder 11, Box 248, PAWA.

2. For example, see Banning, *Airlines of Pan Am since 1927*; Conrad, *Pan Am*; Davies, *Pan Am*; and Josephson, *Empire of the Air*. Notable exceptions to the unquestioned triumphal narrative of Pan Am include recent books such as Bhimull, *Empire in the Air*; and Van Vleck, *Empire of the Air*.

3. Karl Marx in 1857 explained his theory of the annihilation of space by time. He wrote, "While capital must on one side strive to tear down every spatial barrier to intercourse, i.e., to exchange, and conquer the whole earth for its market, it strives on the other side to annihilate this space with time, i.e., to reduce to a minimum the time spent in motion from one place to another." Marx, *Grundrisse*, 539.

4. Agnew, "New Global Economy," 133–34.

5. Crane, *Open Boat and Other Tales*, 3.

6. Ernest Hemingway, "On the Blue Water: A Gulf Stream Letter," *Esquire*, April 1936, 31–32.

7. Zuelow, *History of Modern Tourism*, 58.

8. Sutter, "The Tropics," 182.

9. "Underwater Cultural Heritage," UNESCO, accessed August 1, 2018, http://www.unesco.org/new/en/culture/themes/underwater-cultural-heritage/underwater-cultural-heritage/wrecks/.

10. Blewitt, *Celestial Navigation for Yachtsmen*, 47; Bradford, *Mariner's Dictionary*, 232–33.

11. For a historical overview of shipping in the waters surrounding the United States, see Roland, Bolster, and Keyssar, *Way of the Ship*.

12. "Havana Newspaper Man Describes Plane Trip from That City Here," *Key West Citizen*, January 21, 1928, from Folder 11, Box 248, Collection 341, PAWA.

13. Appadurai, *Modernity at Large*, 37–39.

14. For example, see Walter Kirn's novel *Up in the Air*.

15. "The Logic of the Air," *Fortune*, April 1943, 72–74.

16. In the midst of imperial competition, Pan Am was slower to expand its services to European-controlled islands in the Caribbean. "The French," as one American visitor explained, "were saving the franchise for some French company." Waldo Schmitt to Alexander Wetmore, March 6, 1937, Box 91, Record Unit 7231, SI.

17. Roland, *Model Research*. For example, see Public Law 271, 63rd Cong., 3rd sess., March 3, 1915 (38 Stat. 930) https://history.nasa.gov/SP-4103/app-a.htm #1. For more on U.S. federal funding of scientific and technological development, see Dupree, *Science in the Federal Government*.

18. Van Vleck, *Empire of the Air*, 18–52.
19. Schwartz, *Flying Down to Rio*, 221–58.
20. Bender and Altschul, *Chosen Instrument*.
21. "First Passenger Flight, Key West–Havana, January 16, 1928," Folder 11, Box 248, PAWA.
22. "Panama's Distinctive Stability," *Panama Times*, January 23, 1926.
23. "First Airmail Miami-Cristobal-Miami, 1929," Folder 6, Box 250, PAWA.
24. Banning, *Airlines of Pan Am since 1927*. The old Pan Am air terminal is now Miami's city hall.
25. Van Vleck, *Empire of the Air*, 1–17.
26. Roger Connor, "Amelia Earhart and the Profession of Air Navigation," Smithsonian National Air and Space Museum, February 12, 2013, https://airandspace.si.edu/stories/editorial/amelia-earhart-and-profession-air-navigation.
27. "Profits from Passengers," Folder 11, Box 248, PAWA.
28. Van Vleck, *Empire of the Air*, 246; Peter Dunn, "Why Hasn't Commercial Air Travel Gotten Any Faster since the 1960s?," Ask an Engineer: MIT School of Engineering, February 19, 2009, https://engineering.mit.edu/engage/ask-an-engineer/why-hasnt-commercial-air-travel-gotten-any-faster-since-the-1960s/.
29. Rodrigue and Notteboom, *Geography of Transport Systems*, 249–50.
30. Quoted in Massey, *Space, Place, and Gender*, 154.
31. Gabe Bullard, "The Surprising Origin of the Phrase 'Flyover Country,'" *National Geographic*, March 14, 2016.
32. Benjamin, *One-Way Street and Other Writings*, 52.
33. Schivelbusch, *Railway Journey*; Sutter, *Driven Wild*.
34. Walcott, *The Antilles*, December 7, 1992.
35. "36 Hours," *New York Times*, accessed May 15, 2020, https://www.nytimes.com/column/36-hours. "36 Hours" is a travel series that features various destinations.

Bibliography

Archival Sources

Cleared to Land: The Records of the Pan American World Airways, Inc. Special Collections, University of Miami (PAWA).

Smithsonian Institution Archives, Washington DC (SI).

Published Works

Agnew, John. "The New Global Economy: Time-Space Compression, Geopolitics, and Global Uneven Development." *Journal of World-Systems Research* 7, no. 2 (2001): 133–54.

Appadurai, Arjun. *Modernity at Large: Cultural Dimensions of Globalization*. Minneapolis: University of Minnesota Press, 1996.

Banner, Stuart. *Who Owns the Sky? The Struggle to Control Airspace from the Wright Brothers On*. Cambridge MA: Harvard University Press, 2008.

Banning, Eugene. *Airlines of Pan Am since 1927*. McLean VA: Paladwr Press, 2001.

Bender, Marylin, and Selig Altschul. *The Chosen Instrument: Pan Am, Juan Trippe; The Rise and Fall of an American Entrepreneur*. New York: Simon and Schuster, 1982.

Benjamin, Walter. *One-Way Street and Other Writings*. Translated by J. A. Underwood. New York: Penguin Classics, 2009.

Bhimull, Chandra D. *Empire in the Air: Airline Travel and the African Diaspora*. New York: New York University Press, 2017.

Blewitt, Mary. *Celestial Navigation for Yachtsmen*. London: Adlard Coles Nautical, 1997.

Bradford, Gershom. *The Mariner's Dictionary*. New York: Weathervane Books, 1972.

Covert, Lisa Pinley. *San Miguel de Allende: Mexicans, Foreigners, and the Making of a World Heritage Site*. Lincoln: University of Nebraska Press, 2017.

Conrad, Barnaby. *Pan Am: An Aviation Legend*. Emeryville CA: Woodford Press, 1999.

Crane, Stephen. *The Open Boat and Other Tales of Adventure*. New York: Doubleday & McClure, 1898.

Davies, R. E. G. *Pan Am: An Airline and Its Aircraft*. Twickenham, England: Hamlyn, 1987.

Dupree, A. Hunter. *Science in the Federal Government: A History of Policies and Activities to 1940*. New York: Harper & Row, 1964.

Goetz, A., and Lucy Budd, eds. *The Geography of Air Transport*. Burlington VT: Ashgate, 2014.

Josephson, Matthew. *Empire of the Air: Juan Trippe and the Struggle for World Airways*. New York: Harcourt, Brace, 1944.

Kern, Stephen. *The Culture of Time and Space, 1880–1918*. Cambridge MA: Harvard University Press, 2003.

Kirn, Walter. *Up in the Air: A Novel*. New York: Doubleday, 2001.

Marx, Karl. *Grundrisse: Foundations of the Critique of Political Economy*. Translated by Martin Nicolaus. 1858. New York: Penguin Books, 1973.

Massey, Doreen. *Space, Place, and Gender*. Minneapolis: University of Minnesota Press, 1994.

Merrill, Dennis. *Negotiating Paradise: U.S. Tourism and Empire in Twentieth Century Latin America*. Chapel Hill: University of North Carolina Press, 2009.

Morrow Jr., John H. *The Great War in the Air: Military Aviation from 1909 to 1921*. Tuscaloosa: University of Alabama Press, 1993.

Rodrigue, J., and Theo Notteboom. *The Geography of Transport Systems*. New York: Routledge, 2017.

Roland, Alex. *Model Research: The National Advisory Committee for Aeronautics, 1915–1958*. Washington DC: NASA, 1985.

Roland, A. W., W. Jeffrey Bolster, and Alexander Keyssar. *The Way of the Ship: America's Maritime History Reenvisioned, 1600–2000*. Hoboken NJ: John Wiley & Sons, 2008.

Schivelbusch, Wolfgang. *The Railway Journey: The Industrialization of Time and Space in the Nineteenth Century*. Oakland: University of California Press, 2014.

Schwartz, Rosalie. *Flying Down to Rio: Hollywood, Tourists, and Yankee Clippers*. College Station: Texas A&M University Press, 2004.

Sheller, Mimi. *Aluminum Dreams: The Making of Light Modernity*. Cambridge MA: MIT Press, 2014.

Soluri, John. "Empire's Footprint: The Ecological Dimensions of a Consumers' Republic." OAH *Magazine of History* 25, no. 4 (2011): 15–20.

Sutter, Paul S. *Driven Wild: How the Fight against Automobiles Launched the Modern Wilderness Movement*. Seattle: University of Washington Press, 2002.

———. "The Tropics: A Brief History of an Environmental Imaginary." In *The Oxford Handbook of Environmental History*, edited by Andrew Isenberg, 178–204. New York: Oxford University Press, 2014.

Van Vleck, Jennifer. *Empire of the Air: Aviation and the American Ascendancy*. Cambridge MA: Harvard University Press, 2013.

Walcott, Derek. *The Antilles: Fragments of Epic Memory; The Nobel Lecture*. New York: Farrar, Straus, and Giroux, 1993. https://www.nobelprize.org/prizes/literature/1992/walcott/lecture/.

Yano, Christine R. *Airborne Dreams: "Nisei" Stewardesses and Pan American World Airways*. Durham NC: Duke University Press, 2011.

Zuelow, Eric. *A History of Modern Tourism*. New York: Palgrave, 2016.

4

From the "Romance of Industry" to the "National Soul"

Promoting Travel in the Pan American Union

ANADELIA ROMO

In 1889 representatives from across Latin America and the United States joined together at the First International American Congress, thus forming the precursor to the Pan American Union.¹ Organizers sought to foster diplomatic exchange and trade relations not only through predictable conference meetings but also with a remarkable six-week luxury rail tour for Latin American dignitaries. As the tour made its slow progress through manufacturing sites and major cities of the Eastern Seaboard, participants found themselves showered with unusual souvenirs, such as the handkerchiefs embroidered with the faces of American presidents distributed at one Massachusetts mill.² By the end of the journey, despite such questionable keepsakes, the effort to draw the Americas closer through travel had met with success, at least according to one enthusiastic press report. As a delegate from Venezuela declared to U.S. reporters, the "trip had shown that all previous opinions regarding blood and language being barriers to perfect understanding between nations were wrong."³ Meanwhile, Chile's delegate observed that connections developed during the tour could help increase trade with the United States.⁴ As the expedition reveals, travel was critical to the goals of the Pan American Union from its very beginnings.

Despite its professed intentions to promote mutual understanding across the Americas, the Pan American Union (PAU) has been dismissed by many as a crude tool of U.S. imperialism.[5] Given the location of headquarters in Washington, the early dominance of U.S. directors, and a budget provided in large proportion by U.S. contributions, such an assessment certainly has its merits.[6] More recent work, however, has focused on the cultural efforts of the PAU, an arena that complicates the picture of U.S. dominance, given the often significant investment by Latin Americans themselves.[7] These political and cultural realms were by no means completely separate, however, nor were imperialistic efforts always incompatible with Latin American agendas. With this background in mind, this chapter turns to an analysis of the changing role of travel and tourism within the organization from its origins to the 1960s and demonstrates how travel became central to parties on all sides, revealing both the imperialistic aims undertaken by U.S. actors as well as assertive national agendas promoted by various Latin American interests.

The Good Neighbor era from the Great Depression to the end of World War II reveals this dynamic in particular relief and stands apart from earlier and later periods. Before this juncture the organization had promoted travel to Latin America primarily as a way for U.S. investors to view firsthand the resources the region had to offer. This commercial focus would resurface again in the 1950s and 1960s, only slightly updated with the language of developmentalism. The 1930s and 1940s, in contrast, marked an era distinctive for its promotion of travel in cultural terms. Here U.S. and Latin American needs and interests came together in sometimes surprising ways as travel became, at least for a short while, more distant from the commercial aims that had so long guided the organization. Indeed, travel during this time gained billing as an almost utopic pathway to cultural exchange and greater hemispheric understanding. Most of all, shaped especially by Latin American nationalistic currents that granted new priority to ethnic and folk cultures, promoters both within and outside the region agreed increasingly that travel offered firsthand access to Latin America's "authentic" cultures—believed to be located in folk, popular, and indigenous traditions.[8]

The Romance of Industry: Early Travel in the PAU

While the PAU is often identified as a critical proponent of inter-American travel, historians have not yet sufficiently focused on this important aspect of the organization.[9] This chapter grants new attention to PAU travel initiatives, from the origins of the organization to its transition to the OAS in the 1960s, and examines the way in which promotion of travel shifted over this time, with a particular focus on the 1930s and 1940s. Although the archival record for the institution is sparse, the printed record is exceptionally rich. For my evidence, I draw upon travel guides produced by the PAU as well as the *Bulletin of the Pan American Union* (BPAU), the official organ of the organization. Scholars have noted that by the 1920s there was an explosion of information about Latin America in the United States, much of it produced by a reenergized publication team at the PAU.[10] Yet, as I show, important efforts in travel predated these works, and some of the PAU's most interesting efforts came later. In particular, I call attention to the PAU Travel Division, a little-known department created in 1933, as well as to the proceedings of the new Inter-American Travel Congresses. Throughout, I have restricted this study to the context in which the bulk of the action and attention was dedicated: the English-language texts that urged U.S. citizens to understand, explore, and exploit Latin America. Although the PAU published select travel works in Spanish and although the BPAU was published in Spanish and Portuguese, travel promotion as a whole was overwhelmingly conducted in English, calling attention to the fact that these early inter-American exchanges were imagined to flow primarily in one direction.[11]

Due to gaps in the archival bureaucratic record for the PAU, the actors shaping this larger drive for tourism are not always apparent, but the available evidence certainly reflects both U.S. and Latin American efforts. While authors working in the DC office appear to have drafted most of the material for the BPAU, many of these staff members had Latin American connections. Within this group we can see especially the energy of José Tercero, the early director of the PAU's first Travel Division, created in 1933. Mexican born and educated, he was undoubtedly influenced by the indigenism of the Mexican state in the postrevolutionary era and aware of the larger tourist promotion within Mexico that had begun to highlight popular festivals and indigenous cultures.[12] The next director, Francisco J. Hernández, from Puerto Rico, penned fewer articles in the *BPAU* but was responsible for orchestrating the publication of the heavily "folkloric" guides of the 1940s. Beyond these internal agents, occasionally others across Latin America played a central role. In several significant cases the PAU translated and republished articles from the Latin American press, thus allowing these voices a much broader circulation. And other actors included the Latin American representatives at the PAU's Inter-American Travel Congresses, held first in 1939 and then at fairly steady intervals thereafter (though disrupted initially by World War II). As a whole, the promotion of travel and tourism within the association called upon protagonists from a variety of origins across the Americas.

The official organ of the early Pan American Union (then the International Union of American Republics) gained life four years after its first congress. It began in 1893 with a series of short volumes focused on individual commodities and trade prospects in select areas of Latin America. By 1895 what would become the *Bulletin of the Pan American Union* had settled into a dry compendium of trade statistics organized by country. Dutifully, it reported on tariff policies, industrial indicators, steamship tonnage, and products of a given region, including careful statistics on topics such as the frozen lamb and mutton exports of Argentina in 1896.[13] The focus here was on prospective trade between equals, and thus Latin America was generally portrayed in positive terms, with an emphasis on progress.

This format appears to have been reinvented significantly in 1908–9, when the *BPAU* dedicated itself to new feature articles of greater length and devoted some of them to cultural topics, such as museums across the Americas, or to offerings in the ongoing series Flags and Holidays of the American Republics. According to the editor, these innovations signaled "efforts to make the *Bulletin* not only practical and dignified but an attractive and pleasing agency for educating and informing the world in regard to the progress, commerce, and development of the American republics."[14] Perhaps not coincidentally, the BPAU dedicated new attention to the question of travel in Latin America at this same time. In 1909 the editor reported that many recent correspondents had credited the propaganda of the BPAU for shifting their travel itinerary from the "Orient," or Europe, to South America.[15] Although the cultural scope of the BPAU had been significantly widened, the focus of travel articles at this time remained targeted at the potential investor, with typical articles entitled "Commercial Travelers in Latin America." In a similar vein, a series of submissions by an anonymous author, calling himself simply "Viajero," or Traveler, touted the benefits of travel to Central America with a very practical bent. Advocating that readers venture beyond well-worn tourist routes, he advised a more difficult agenda that instead granted exposure to "the very portions of a country most open to opportunity for the introduction of American goods."[16]

Beyond the work in the BPAU, other publications of the PAU took an even more narrow approach to promoting commerce. To begin with, the organization produced no travel literature at all, but at least several countries—Honduras (1904), Cuba (1905), and Mexico (1911)—merited their own early volumes promoting trade and investment opportunities.[17] It appears that by 1911 the PAU had begun publication of a series of pamphlets for each country, but, as the title for the series—General Descriptive Data and Commerce—indicated, their purpose was to encourage investment, not draw in armchair readers or tourists. Indeed, the guide for Cuba in 1911, for example, contained no information about culture, society, or demographics but offered instead a staid political overview and extensive import and export statistics.[18]

With the outbreak of war in Europe in 1914, the PAU's director observed that the conflict had opened new opportunity for commerce in Latin America: as Europe—land of the Grand Tour for European and U.S. tourists—closed for tourism, travelers could redirect their itineraries to Latin America.[19] This diversion of travel from Europe gained new possibilities with the opening of the Panama Canal, and a new steamship line began serving the Pacific coast, a development the BPAU had already promoted. Travelers from the eastern coasts of the Americas could now gain access to countries such as Argentina in half the time.[20] In response to this new opportunity, the BPAU began to reprint its travel articles in stand-alone form in 1917; these were the first works to break out of the "descriptive data" mold and were eventually incorporated into independent series. The burgeoning Commodities series, an eclectic collection of short pamphlets promoting products such as alpacas, asphalt, and chicle, was now joined by the series American Cities in 1925, while American Nation began in 1930.[21] These lists and others dedicated to travel and tourism expanded, and by the mid-1950s the PAU had coverage that extended to all of Latin America.[22] Moreover, many of these works were updated annually, an undertaking that gives some indication of the scale and energy that the organization devoted to promoting travel.

Changes also followed from energetic efforts of individual authors within the PAU. Such was the case with Henry Alfred Reid, who penned many of the early travel works. Officially listed as trade advisor to the PAU, he also oversaw the reorganization of the PAU publications program after 1909, when he was hired away from the U.S. Department of Commerce.[23] A prolific writer and editor, Reid titled a 1914 article "As Tourist and Trader See Costa Rica," which expressed well his overall focus.[24] This emphasis on the practical rather than romantic was the keystone of his approach, one that he worked to apply to all of the publications. Although Reid certainly found much to be optimistic about, his sense of marvel was awakened equally by the majesty of Mexico's pyramids as by port improvements in Brazil.[25]

Moving into the 1920s, the BPAU defined itself as "a careful record of Pan American progress" and brought together a curious mixture of cultural and commercial promotion, though it continued to lean more heavily toward the latter.[26] A typical issue might have a discussion of Incan weavings interspersed with other contributions that detailed telephone lines in Mexico, industrial potential in Chile, and agricultural efforts in Argentina.[27] Ostensibly, all of these articles were intended to promote travel and exchange, but those dedicated directly to the leisure tourist were rare and numbered only a few per year. More common were travel essays that instead stressed the resources available to investors and potential immigrants. One writer rhapsodized about Colombia, emphasizing the seduction of profits and stating that there were "few other places more suggestive of hidden resources or more open to the romance of industry."[28]

Gradually, however, there emerged a shift in the BPAU's emphasis, as staffers began to consider the economic potential of tourism itself. A pioneering 1927 article, dedicated to praising tourist guides, explained to readers that "this is the age of travel, and the tourist crop, assiduously cultivated, sometimes yields huge returns."[29] As its author concluded, if all of the countries of Latin America had well-crafted guides, the "$136,000,000 which American tourists now annually spend in Europe would be diverted southward, for in point of natural beauty and human picturesqueness few countries of the world excel [sic] those of Latin America."[30]

The picturesque, however, was rarely taken up as the centerpiece of the guides themselves. A prime example of this was the guide for Rio de Janeiro, first published in 1917, which took considerable pains to emphasize the city's progress, modernity, and "civic achievement," attempting to correct the impression that Rio was a tropical place of beauty alone. Instead, one writer insisted, "For once the city should be considered from the viewpoint of economic utility and as an example of efficient political government and a pronounced civic success."[31] Likewise, *Seeing South America* in 1927, the first volume in the new Sightseeing series, was similarly distinctive for its decidedly commonsense view.[32]

The publications of the PAU promoted travel with growing enthusiasm through the 1910s and 1920s, but these works remained framed in predominantly commercial terms. As a whole, these early guides took little interest in romantic ideas of the tropics, folk culture, or the exotic more broadly. Questions of ethnic or cultural difference—indigenous or African roots—gained no treatment in the rush to portray the region as understandable, approachable, and equal to the United States in every respect. This treatment changed dramatically beginning in the 1930s and 1940s, a trend introduced in part by the advent of the Good Neighbor policy.

Tourism for Spiritual Communion:
Folk Culture and the National Soul in the 1930s and 1940s

Beginning in 1933, U.S. president Franklin Roosevelt's Good Neighbor policy played a central role in pan-Americanism. The policy, initiated early that year, and the accompanying decline in U.S. military activities in the region granted room for expanding relations between Latin American countries and the United States. The brutality of World War I had brought into question Europe as an ideal of civilization, and it also ushered in the new dominance of the United States as a global investor, especially in Latin America. As World War II unfolded, furthermore, the diplomatic need for regional alliances gave heightened urgency to cultural understanding and exchange across the region, creating a veritable flood of activity across U.S. government agencies and private foundations alike.[33]

Travel publications during the 1930s had been steadily issuing from within the PAU, but the numbers reached new levels with wartime pressures. By 1943, for example, the PAU had devised a curious solution to extend the metaphor of travel in the midst of war: it urged the creation of local U.S. social clubs, which would follow a program crafted and published by the PAU: *Documentary Material for the Good Neighbor Tour: An Imaginary Visit to the Republics of Latin America*.[34] Participants would punch their own virtual steamship tickets on a visit across the Americas, listening to lists of recommended folk music, and sampling food pairings at monthly meetings in their own homes. It was in this wartime period that a new focus on Latin American culture developed, thus departing substantially from the longtime commercial focus of the PAU. To understand this departure, we must examine the priority given to culture in the new PAU Travel Division, as well as in the Inter-American Travel Congress gatherings the division began holding in 1939.

The Travel Division had its origins in late 1933 and began work the next year, under the leadership of José Tercero.[35] The first reports of the division stressed a cultural mission of making Latin America more appealing to U.S. travelers and more widely understood. As Tercero established, "Many observers agree that the great tide of American travel has definitely turned southwards.... Having as one of its paramount purposes the promotion of closer acquaintance and better understanding among the peoples of the New World, the Union is well aware of the importance of travel in achieving these ends."[36] According to Tercero, the PAU had long since begun these efforts, but now, with the Travel Division, there was greater support for them.

This cultural mission for mutual understanding by no means indicated that economic concerns were absent, either within the travel promotion by the PAU or within the larger agenda developed by U.S. government officials. By the late 1930s the U.S. Department of Commerce had become well aware that the number of U.S. citizens traveling to Latin America was exponentially higher than the reverse, and it began using the new Travel Division to gather statistics on travel numbers and expenditures. The resulting report concluded that the potential for tourism had not been recognized, either as a "medium for the promotion of friendly cultural relations, or as the important industry that it is—a very definite factor in the international balance of payments."[37] Other U.S. government offices similarly used the PAU to apply pressure for measures that would yield economic benefits from increased travel. The U.S. secretary of state, for instance, instructed U.S. delegates to the PAU conference of 1938 to be active in pushing tourism within the program.[38] As he advocated, "The development of greater tourist travel is a subject which is of prime importance at the present time to American shipping interests.... The facilitation of tourist travel between the American states is a subject of considerable interest to the Department."[39] Such language indicated new efforts by the U.S. government to encourage travel as part of its economic and diplomatic agenda.

The way this agenda unfolded within the PAU in the 1930s, however, gave little attention to such commercial rationales, even though they had long dominated its own publications. Instead, the new travel agenda of this period revolved around the idea that travel offered a unique window into a culture, a notion that Tercero promoted at the very founding of the Travel Division. Moreover, this culture was increasingly portrayed in ways that gave special attention to folk and indigenous cultures. This focus mirrored an interest in folklore that had been building in Latin America during the twentieth century and that had become a critical part of the reimagined nationalism of the era. As elites and intellectuals tried to move away from slavish imitation of European culture, long represented as the ideal, they looked to native inspiration on their own soil. One of the most studied cases of this trend, and perhaps one of the strongest, was that of Mexico, where elites, artists, and the state itself embraced a new vision of the nation as centered in its indigenous culture. Brazil too represented an important move in this same direction, as modernist thinkers involved in São Paulo's Modern Art Week of 1922 began to theorize about a national identity built on African and indigenous roots, one that would digest outside influences on its own terms rather than imitate them wholesale. These trends spread across Latin America and, though varied in their timing and focus, were consistent in marking the first half of the twentieth century as transformative.[40]

In fact, an interest in culture that reached beyond European prototypes had been a part of the Travel Division from the start. An early report from 1934 noted proudly that American Express had used the division's materials to put together flyers on Carnival in Rio de Janeiro.[41] This interest had grown significantly across the PAU by the end of the 1930s, revealing an increasing focus on folk culture. The BPAU volume of 1937, for example, collected folktales and folk songs from different countries and then published the works as a separate volume later that year.[42] In 1939 the PAU publication of *Some Latin American Festivals and Folk Dances* continued the trend. Subsequent years brought more regional efforts along these lines, such as one dedicated to the festivals of Central America and Panama (1940) and another to Mexico (1943).[43]

It is worth distinguishing among these different publications to examine their intellectual origins, and the 1939 publication on festivals and folk dance is a particularly interesting example. The work compiled four articles published in BPAU in 1939. Two were authored by editorial staff at the PAU, and two were edited and reprinted from periodicals in Mexico and Chile. To start with, the contributions from the PAU both stressed the importance of folk culture in dramatic terms. Heitor Bastos Tigre opened his article on Carnival in Brazil emphatically, declaring, "Nothing reflects the soul and temperament of a nation better than its folk customs. They are a faithful mirror of its real nature." Carnival itself was defined as a "popular festival par excellence," the "most representative of all Brazilian folk festivals."[44] This language matched well with the populist language of the Vargas era in Brazil at this time. Intellectuals and state officials there also sought to prioritize an "authentic" national culture of "Brazilianness," or *brasilidade*, drawing heavily on what were deemed popular and traditional practices.[45]

In the second contribution by PAU staff, Francisco J. Hernández, head of the Travel Division, detailed indigenous festivals in Peru, where "old customs" of celebration were a topic of growing interest: "The number of travelers interested in the Indian lore typical of many Latin American countries increases steadily as information is circulated in the United States regarding the fiestas, fairs and many indigenous or hybrid ceremonies which are to be found off the beaten track in the heart of the countries to the south."[46] After a brief overview relying on the authority of a "well-known folklorist," Hernández went on to summarize the work of Peruvian intellectual José Uriel García (most likely the well-known folklorist referred to earlier). García was a socialist thinker and leader in the New Indian movement of Cusco who had sought since the late 1920s to valorize indigenous culture and folklore in Peru.[47] His work treated the celebrations with respect and focused in particular on the connections between contemporary festivals and their Incan predecessors. This same interest came through in the summary by Hernández, who wrote that "many of the old native customs still persist, in only slightly altered form."[48]

Latin American nationalist thinking emerged more directly in an article the same year on Mexican folk dances, reprinted from the periodical *Mexican Art and Life*.[49] With only a brief run (1938–39), this magazine was issued by Mexico's Autonomous Department of Press and Publicity (Departamento Autonomo de Prensa e Publicidad, or DAPP), a federal propaganda office created by the populist president Lázaro Cárdenas. Perhaps predictably, the content and the covers of this publication highlighted folk and indigenous elements, reinforcing the visual culture promoted by the Mexican state during this time. Folk dance also gained attention in the new definitions of national culture, and it was this topic that formed the heart of the article. According to the author, the dances described derived from two "racial stocks," the "Aboriginal" and the Spanish, but this author's interpretation granted priority to the former, claiming that "the manifestations closely related to native traditions are the most interesting."[50] For authors like this one, indigenous culture represented a critical part of national culture, and practices were praised according to their fidelity to native traditions.

Finally, the last article of the four, "Some Latin American Festivals: Folk Dances of Spanish America," was "based chiefly" on the work of Peruvian author Luis Alberto Sánchez. Sánchez, exiled from Peru, had taken up residence in Chile and published his work in the official organ of Chile's Ministry of Education, a journal then under the direction of a Chilean folklorist who was an advocate of the popular arts.[51] This article treated a variety of dances from Mexico and South America, with a focus on the narratives and local characters that regional dances revealed. Again, dance here was seen to be central to national identity, although indigenous roots per se gained less attention. As Sánchez declared emphatically, "If there is any popular art that springs directly from the emotions and passions of the people, it is dancing."[52] As this collection of four articles from the *Bulletin of the Pan American Union* suggests, the PAU publications for tourists during the late 1930s drew considerably on intellectual and cultural trends in Latin America, in many cases using scholarly production from the region directly, while in others filtering it and translating it for a broader English-speaking audience.

The Pan American Union prioritized this indigenous focus and interest more broadly as well. It is no coincidence that the PAU hosted

13. Camilo Blas, *Peruvian Country Dance*. Sketch printed in *Bulletin of the Pan American Union*.

the first Inter-American Indian Congress in 1940 and established the new Inter-American Indian Institute in this same moment—although at its founding that organization, pledging to "state and solve" the "Indian problem," appeared to be more interested in modernization than in matters of folklore.[53] Perhaps most remarkable was a Pan American Union meeting called in 1943 to produce an official common resolution on folklore. This gathering of state and cultural officials defined folklore as "spiritual manifestations of the people" and as being in danger "from superficial, strange and destructive influences."[54] Some years later the Pan American Institute of Geography and History (established in Mexico in 1928) formed a Comité de Folklore in 1953 and would publish the journal *Folklore Americano* in Peru beginning the same year.[55] Although these efforts found official support in the PAU, they had significant roots in a broader Latin American interest in indigenous culture and, by the 1940s, a larger interest in folklore more broadly.

The focus on folklore was especially apparent in the Music Division of the PAU, founded in 1940. In 1942 it published a collection of popular and folk songs, the *Cancionero Pan Americano* (Pan American songbook) and had undertaken significant effort to collect and document folk music traditions across Latin America.⁵⁶ These efforts were possible in part because Latin American musicologists and intellectuals had already come to embrace the study of folk culture and folk music as a basis for national identity. As a result, they were often inclined to view U.S. efforts to build up these studies, as well as inter-American networks for their support, with interest. In the Music Division especially, "given the preexisting interest in folklore across the hemisphere, the new U.S. cultural policies reinforced a transnational community of scholars." Far from imposing a foreign concern onto the region, the PAU's music agenda showed considerable influence from Latin American actors already long active and already in place.⁵⁷

In contrast to these celebrations of indigenous or folk influences, the incorporation of African roots or Afro-Latin populations took longer. Even as Brazil officially granted support to ideas of racial mixture and Cuba ushered in an Afro-Cubanismo movement in the 1930s, such movements were often limited in their full acceptance of blackness, even at the symbolic level.⁵⁸ Such hesitations were reflected in the PAU literature as well. Brazil's guide of 1937, for example, stressed that the population was "for the most part of Portuguese descent" and rushed to assure readers that European immigration, though long "retarded" by slavery, had picked up with abolition in 1888.⁵⁹ The language here, though clearly standardized by PAU editors, seemed taken directly from the racism of Brazilian writers of the time. Indigenous cultural practices remained less controversial thanks to a longer interest in *indigenismo* that had developed in various forms across Latin America and had succeeded in gaining official endorsement in areas like Mexico and Peru.

Indeed, many Latin American countries by the 1930s were attempting to develop their own tourism by playing up indigenous roots or regional folklore for domestic and international visitors alike. Perhaps the clearest example of this pattern is Mexico, where scholars have noted that ideas about folkloric authenticity were enthusiastically seized as prime material for the tourist industry during this time.⁶⁰ In cities such as Pátzcuaro and Veracruz, regional and festive traditions were mobilized in order to encourage citizens to travel within their country's borders throughout the 1930s. Similar processes took place across Latin America, whether in the promotion of indigeneity in Cusco or the crafting of a folk culture in Argentina's Northwest. In a later development the city of Salvador in northeastern Brazil marshaled ideas of blackness to set itself apart in the growing domestic travel scene of the 1950s.[61] There, in the absence of indigenous folk traditions to draw upon, promoters turned to Afro-Brazilian traditions that by the 1950s were finally gaining broader acceptance in cultural terms.

Yet within the travel publications of the PAU, it is worth noting that there was no one consistent voice. In a somewhat different vein from the scholarly approach to folk culture as essential to the nation, a later mimeographed guide to holidays in Mexico struck a somewhat bemused tone, declaring, "In Mexico it's always fiesta time." This guide was penned by PAU staffer Elizabeth Hastings, most likely a U.S. native.[62] Instead of a scholarly approach, this publication took a more informal and chatty tone and was more directly intended for the general tourist. Here the tourist was simultaneously instructed how to see the most authentic festivals outside the capital and given cultural lessons that stressed the exotic and the foreign nature of their hosts. The reader was advised that one could ostensibly follow the Catholic calendar for the festivals but warned that it might be disrupted by the figure of Don José, a fictitious persona used to show the many intervening personal motivations that might affect or change the schedule. As the guide quipped, "How will one know when the festival will begin—'Quien sabe—who knows?'" At the end of the visit, Hastings concludes, "The visitor boards his train or steps into his car and returns to the capital, thrilled and exhilarated by this new glimpse into the pagan heart of mysterious Mexico."[63] Mexico here was exotic, and impossible to understand fully, but the tourist was encouraged to go and observe this difference directly.[64]

The Inter-American Travel Conferences

Within these larger efforts to promote travel during this time, the Travel Division focused especially on organizing the Inter-American

Travel Congress, hosted first in San Francisco in 1939 and intended to "strengthen social and economic ties."[65] The next meeting was held two years later, in Mexico, and though the following one was delayed due to World War II, after the third conference (1949), meetings continued to be held roughly every two years.[66] The meeting was the first in which the PAU explicitly addressed tourism, although others, such as the Pan American Highway Congress (first held 1925) and the Commercial Aviation Congress (first held 1927), had addressed the mechanics of travel. With some of the transportation logistics already under consideration elsewhere, the first Inter-American Travel Congress granted exceptional priority to exploring how to immerse the traveler in the culture of Latin America and the benefits that such experiences would bring.

Indeed, in the first meeting the potential for travel to bring people together was paramount. One of the final resolutions proclaimed that "common interests will be the more ardently defended as the people who possess them know one another better. This knowledge, however, should be obtained directly from the observation of reality, rejecting anything esoteric which could obscure that reality.... The common interests of all Americans require that travel be considered in its higher sense as a means to achieve a closer cultural and spiritual rapprochement of the peoples of the Americas."[67] The idea of avoiding the "esoteric" hints that Latin American representatives may have hoped that travel would correct misrepresentations of the region that people in the United States might harbor. Hollywood had helped create a "vogue" for things Latin American by the 1930s, but its films often stereotyped the region with offensive representations that created ill will among Latin American audience members.[68] On the other hand, U.S. tourists and marketers sometimes took an interest in African-inspired culture with which Latin American elites were still uncomfortable. Such had been the case with the Hollywood film *Rumba*, for example, which portrayed Afro-Cuban dancing in ways that prompted objections from the white elite; the film was banned in Cuba in 1935.[69]

An examination of the proceedings of the first nine conferences, spanning from 1939 to 1965, reveals that the first meeting in particular was dominated by interest in folk and popular cultures. Later years showed a marked difference, with interest in folk culture already declining rapidly in importance by the second congress, in 1941.[70] Yet in the early years of the Travel Division, director José Tercero seems to have shaped the first 1939 congress around the interests of the moment, often expressed in utopic terms. According to the resolutions of the first congress, the goal for travelers was not only to "enjoy the natural beauties and scenic and other attractions . . . but also to acquire a basic understanding of and basic appreciation for historical traditions, cultural heritage, native customs and other expressions of the national or local soul."[71]

For the expression of this soul, the first Inter-American Travel Congress (1939) advocated exposure to festivities and popular cultural practices. National travel boards, responsible for increasing domestic and international travel, were charged with mounting an "inventory of the travel assets" of their homeland, with attention to "special features typical or peculiar to each center," such as "fiestas, celebrations and fairs, whether popular, regional, indigenous, religious, patriotic, etc."[72] In this vein the congress concluded by recommending "the preservation of folk customs, arts, and industries" and the creation of a calendar of folk festivals to be held, as well as special folkloric events in the cities where each travel congress would occur. Congress organizers also recommended the "restoration and protection of archeological, colonial, historical and natural monuments; and the establishment of national parks."[73] The folk cultures of the Americas, then, were imagined not only as precious national resources that included indigenous practice but also as key tourist attractions in need of further promotion and protection.

It should be noted that commercial interests were also invested in this promotion of native traditions and festivals. U.S. travel agents and transport representatives attended the Inter-American Travel Congresses and seem to have shaped certain resolutions. Likewise, their Mexican counterparts showed up in great force when Mexico hosted the conference in 1941. In the BPAU Peruvian folklore article many of the photos were credited to Grace Line, the U.S. cruise ship company connecting New York to the west coast of South America. And the depiction of a Mexican charro in the article chronicling folk dance was provided by National Railways of Mexico, only recently nationalized by President Cárdenas. These images were counterposed by other illustrations and photos by Latin American artists, including some by important figures like the Peruvian indigenist artist Camilo Blas.[74] Here the PAU served to publicize and disseminate ideas about folklore and indigenous culture already developed by intellectuals in Latin America—ideas that were becoming increasingly attractive to other commercial agents in promoting the region as well. Rather than creating its own independent narrative, the PAU drew upon strains of thinking already in place among both Latin American nationalists and business interests across the Americas, reinforcing them both. This narrative, however, would shift again in the postwar era.

Which Roots? Postwar Transitions

The PAU, reorganized as the OAS in 1948, continued to emphasize travel in the postwar era. The director general, former Colombian president Alberto Lleras Camargo, stressed in his annual report of 1948 that "all nations of the western hemisphere have recognized the importance of tourist travel as a factor in national economy, as a cultural force, and as a means of obtaining a clearer understanding of neighboring countries."[75] Despite such grand statements, the larger passion for promoting travel within the organization began to decline, perhaps due to postwar organizational and diplomatic shifts. Chief among these was surely an administrative reorganization in 1947 that, while seeking to centralize the functions of the PAU, simultaneously sapped the energy that had built within the Travel Division by eliminating the division altogether.[76] More broadly, as the United States turned its focus to rebuilding Europe and to the Cold War, priority given to cultural approximation between the United States and Latin America lessened: the urgent needs of wartime cultural diplomacy were no longer pressing. And finally, the shifts seem to have been a result of changing national priorities and cultural trends within both Latin America and the United States, where interest in the folk and indigenous had waned. By the postwar era much of the region was feeling its national identity to be more firmly consolidated, and folk culture became less interesting to state officials than modernization and development.

One of the largest shifts was the extinction of the *Bulletin of the Pan American Union* itself, which was replaced by a glossier journal, entitled *Américas*, in 1949. This journal catered more clearly to the general U.S. public and maintained some of its previous cultural agenda, with regular features such as a Latin American discography. Popular festivals too continued to gain attention, with a special calendar in each issue, and a full issue dedicated to tourism appeared each year. Although many of its readers wrote in approvingly, its somewhat eclectic format elicited the disdain of one reader, who claimed it would not interest any Brazilians because it "lacks character. It is and it is not a travel publication, a fruit salad, a mixture of politics, diplomacy, propaganda. . . . It looks like an academic magazine written by scholars and it seems half dead. . . . It hurts to see an ideal going down the drain."[77]

The new cultural agenda of the magazine seems to have sapped attention from the topic elsewhere. By 1949, at the third travel congress, the language had already changed so that the resolutions on "cultural aspects" of travel promotion were minimal, having been reduced to less than half the space dedicated to sports. Tourism was no longer seen as a tool to understand the national soul, although it continued to garner grand language describing it as the surest path to "strengthening solidarity among the peoples of the Americas."[78] And a significant blow was delivered at the fifth travel congress, in 1952, when the United States finally admitted, somewhat tersely, that a standard approach to simplifying tourist entry requirements would not be considered "in the foreseeable future," thus putting an end to fraternal ideals of freer travel across the Americas.[79] By the early 1960s the travel congress had adopted much of the developmentalist bent of the OAS itself, calling for increased funds from the Alliance for Progress and the Inter-American Development Bank for hotel construction, as well as urging in 1965 that tourism be understood "not only as services but also as activities of basic interest to economic development."[80]

Changes were slower to appear in the guides themselves, which at first glance seemed to continue much as before. The OAS initially maintained the Travel in the Americas series, dedicated to helping tourists plan leisure vacations in Latin America.[81] Yet in these years the organization generally stepped away from granting folk culture any real priority, thus marking a new tone. The Cuba guide for 1957, for instance, written by Paula Armstrong, of the OAS's Editorial Division, opened by describing the country as "a gay happy land with a perpetual holiday air" that marked the country as festive and exotic for fun-seeking vacationers. Overall, the author granted little attention to popular culture or folk interests, except briefly in the realm of evening entertainment. There Armstrong advised that rumba was best viewed not in the nightclubs but in "native dance halls," where one could see "dusky dancers in their natural habitat," a questionable phrasing that veered uncomfortably toward linking Afro-Cubans to the animal world.[82] This language emphasized otherness rather than familiarity and made Cuban dance not a way into the Cuban soul or a path to greater understanding but an exotic display.[83] The Travel in the Americas series ended for unknown reasons in 1959, perhaps as collateral damage in the larger diplomatic fallout that accompanied the Cuban Revolution.

Whether as a result of the events in Cuba or not, by the 1960s the OAS had discontinued its earlier guides directed at tourism. Instead, it retook its older American Nations volumes and folded them together into a new series, the American Republics. Now published by the Department of Information and Public Affairs, these publications targeted an audience that seemed to be less tourist than potential investor. In many ways this format returned to that of the early twentieth century: it was heavy on investment information, with only a smattering of culture. Information for the traveler had been reduced from the amounts offered in guides of the previous decade, and tables of exports, imports, and principal trading partners filled the last pages. As a symbol of an earlier era, however, a small paragraph labeled "Music and Folklore" lingered in each volume.[84]

Conclusion

As a whole, the Pan American Union has followed the lines of U.S. diplomacy so closely that it is tempting to see it as a mere puppet of a larger U.S. agenda. Yet to see the actions of the PAU as unilateral disregards the fact that Latin Americans also wished to promote their own modernity at the turn of the century and that later they too turned to developmental language to describe their reality. Most critically, as we have seen, while the wartime effort to encourage tourism through folk culture was surely driven by U.S. interests, it was also embraced by Latin American states and their elite, who hoped to reorient their economies after the 1930 collapse and gain new endorsement for their own efforts at nation building. In tourism, then, we may see the union of imperialistic agendas from the United States as well as nationalistic agendas in Latin America.

Travel itself was certainly seen as a positive by Latin American diplomats and officials engaged with the PAU; they promoted and endorsed travel as a tool for their own benefit. Although the PAU undoubtedly often promoted U.S. interests, it also reflected the interests of the Latin Americans who worked for the organization, the diplomats and officials who submitted information to its publications and attended its conferences, and the visions of Latin American authors, who were given a broader audience. And for a brief time the Pan American Union revealed unification across borders in surprising ways as consensus formed that travel, as well as exposure to "authentic" folk cultures, represented the ideal way to understand nations and each other.

Notes

1. The organization commonly designated as the Pan American Union changed both its name and its mission various times. The organization began in 1889 as the International Union of American Republics and became known as the Pan American Union only in the first decades of the twentieth century. In 1948 the entity underwent a reorganization to form the Organization of American States, the body that continues to this day. For the purposes of clarity and simplicity,

and following more general usages, I have used Pan American Union or PAU to refer to the organization and its conferences before 1948, and I have used the Organization of American States or OAS to refer it in the post-1948 period.

2. You could choose to blow your nose on President Washington's or President Harrison's portrait, if desired. "Inspecting the Mills: The Trip of the Pan-American Delegates; A Visit to Lawrence and Lowell in Which the Members of the Congress Were Much Interested," *New York Times*, October 8, 1889, 2.

3. "What They Think of Us: Some Impressions Received by Our Pan-American Guests," *New York Times*, November 12, 1889, 5. As historians have noted, the actual results of the congress have been judged more critically, including by many Latin American nations that at the time had little use for the organization or its meetings. Smith, "First Conference of American States," esp. 25–27.

4. "What They Think of Us," 5.

5. Mexican economist Alonso Aguilar traces this interpretation elegantly from a Latin American perspective; see Aguilar, *Pan-Americanism*. Historian Mark T. Berger is similarly critical of U.S. imperialism in the role of the PAU, although the organization is not his exclusive focus. He argues that the professionalization of the study of Latin America closely followed U.S. diplomatic and commercial aims, and he views the PAU as a part of this endeavor; see M. Berger, *Under Northern Eyes*. More recent research having the PAU as its focus has often contained arguments along similar lines. For historian Pablo Palomino, for example, the PAU's Music Division ultimately represented the imperial actions of a dominant power, with their impact being felt by less empowered allies. See Palomino, "Nationalist, Hemispheric, and Global." José Manuel Espinosa provides a largely sympathetic overview of the early congresses and the union but describes the 1928 congress as the lowest point of mutual trust among the member nations due to U.S. military interventions in the region; see Espinosa, *Inter-American Beginnings*, 7–28.

6. Financial support for the union came from contributions based proportionately on population numbers; the United States therefore far outstripped even the next-largest country, Brazil. For the early history of the organization see Smith, "First Conference of American States." Mark Gilderhus takes up the story of Woodrow Wilson's embrace of pan-Americanism in his book *Pan American Visions*; the origin of pan-Americanism is treated more broadly in Fagg, *Pan Americanism*. There is still no one volume that traces the organization critically over its full history, perhaps due to the many roles undertaken by the PAU and its long trajectory through inter-American affairs.

7. Claire Fox, for example, in her study of the influential Visual Arts Section of the PAU, posits considerable intellectual autonomy for the division under its Cuban-born director, José Gómez Sicre, in the late 1940s through the 1960s. Historian Corinne Pernet's study of the PAU's Music Division further argues (in contrast to Palomino, "Nationalist, Hemispheric, and Global") that the division offered new prestige, funding, and opportunities for Latin Americans already

engaged in the promotion of their own musical agendas. In a less cultural vein, Donna Guy proposes that the case of the Pan-American Child Congress reveals an early, important instance of Latin American leadership within the organization. See Fox, *Making Art Panamerican*; Pernet, "For the Genuine Culture of the Americas"; and Guy, "Politics of Pan-American Cooperation." An edited collection dedicated to the PAU (Sheinin, *Beyond the Ideal*) provides an excellent overall survey; several contributions reveal Latin America as an active player in the PAU project in different eras. For the period most relevant to this study, see especially Leonard, "New Pan Americanism." For the role of Latin American cultural interlocutors more broadly in promoting pan-Americanism (although not always the PAU), see Cándida Smith, *Improvised Continent*.

8. It is also important to note that the folklore movement within Latin America was itself tied to the promotion of tourism, often domestically as well as internationally. This point is made particularly well for the Andes in Mendoza's "Tourism, Folklore and the Emergence of Regional and National Identities," as well as her *Creating Our Own*. While the rise of folklore in individual countries in Latin America has been treated by other scholars, most notably in Argentina by Oscar Chamosa, much more remains to be done, and a broader survey of the region as a whole is certainly overdue. Pernet begins this project with a concise but insightful overview of the region as a whole in her study of Latin American folk music within the PAU. Chamosa, *Argentine Folklore Movement*; Pernet, "For the Genuine Culture of the Americas." For the construction of the concept of the folk and authenticity, the best overall study is Bendix, *In Search of Authenticity*.

9. Beyond the works in this volume, historical treatments of tourism in Latin America have been steadily growing in number. For a representative look, see especially Merrill, *Negotiating Paradise*; and D. Berger and Wood, *Holiday in Mexico*.

10. J. Valerie Fifer views these efforts to disseminate information as offshoots of a larger commercial drive by the United States. For the case of the Pan American Union more specifically, see Fifer, *United States Perceptions of Latin America*, 155–58. For a useful contemporary overview of activities and publications of the PAU at that time, see Albes, "Pan American Union."

11. The fact that only one isolated guide was dedicated to travel from south to north speaks also to the fact that this was never the main agenda of the Pan American Union or its travel initiatives; see Pan American Union, *Viajando por los Estados Unidos*. It is worth noting that the tone of this guide was very similar to that of the others prepared for Latin America during this time.

12. It is not clear from Tercero's obituary in the *Bulletin of the Pan American Union* when he moved to the United States, but he began work for the PAU in 1927 in the Editorial Division before going on to head the Translations Department and later the Travel Division. "José Tercero: In Memoriam," *Bulletin of the Pan American Union* 73 (1939): 478–79.

13. See *Bulletin of the Pan American Union* 3 (January 1896): 378. The name of these early volumes was technically the *Bulletin of the International Bureau of the American Republics* until the end of 1910, although they are now often cataloged as numbers of the *Bulletin of the Pan American Union*.

14. "Changes in the Bulletin," *Bulletin of the Pan American Union* 30 (February 1910): 165. These editorial shifts may have been in part due to a new director of the PAU, John Barrett, who took over this role in 1907.

15. "Increase of Travel to Latin America," *Bulletin of the Pan American Union* 28 (January 1909): 5.

16. Viajero, "Travel Notes in Central America," *Bulletin of the Pan American Union* 32 (January 1911): 77.

17. Pan American Union, *List of Publications*. The Cuba guide is curiously not listed in this compendium, perhaps because it was out of print.

18. Pan American Union, *Cuba*.

19. Barrett, *Our Opportunity in Latin America*, 3.

20. "Increase of Travel to Latin America," 5.

21. Because of the many different publication years for individual titles, it is sometimes difficult to tell when a series was initiated. In addition, series sometimes incorporated individual titles that had already been in circulation.

22. An apparently short-lived series, Sightseeing, began in 1927; later, in 1943, the Travel Division created the series Travel in the Americas, which featured individual volumes on different countries.

23. Fifer, *United States Perceptions of Latin America*, 156.

24. To gain some idea of his impact, note that Reid authored half of the thirty volumes in the City series, and he also served as editor for the Nation series.

25. Reid, *Mexico*; Reid, *Glances at Ports and Harbors*.

26. For this phrasing, see, for example, the back cover of Reid, *São Paulo*.

27. See, for this example, the *Bulletin of the Pan American Union* 61 (May 1927).

28. H. A. Caracciolo, "Colombia as Seen by a Tourist," *Bulletin of the Pan American Union* 61 (December 1927): 1213.

29. "Guides and Guidebooks," *Bulletin of the Pan American Union* 61 (January 1927): 17.

30. "Guides and Guidebooks," 27.

31. Albes, *Rio de Janeiro*, 1.

32. Pan American Union, *Seeing South America*.

33. Especially notable in these efforts was the creation of the Office of the Coordinator of Inter-American Affairs (OCIAA) in 1940. In addition, the State Department only two years earlier had created the Cultural Affairs Division, which focused exclusively on Latin America. Perhaps the most visible record of this moment came from the studios of Walt Disney, enlisted to produce Good Neighbor propaganda for the OCIAA. For a case study of these cultural efforts in Brazil, see Tota, *Seduction of Brazil*. For the period in Latin America as a whole, see Drinot and Knight, *Great Depression in Latin America*.

34. Pan American Union, *Documentary Material for the Good Neighbor Tour*.

35. Tercero served until his untimely death in 1939, just after the first PAU travel congress.

36. José Tercero, "Travel in the Americas," *Bulletin of the Pan American Union* 69 (January 1935): 31.

37. Pan American Union, *Inter-American Travel Statistics*, n.p.

38. United States, Department of State, "The Secretary of State to the American Delegation to the Eighth International Conference of American States, Washington [undated]," 68, Foreign Relations of the United States Diplomatic Papers, 1938, The American Republics, volume 5 (Document 55), 710.H/215, accessed July 15, 2018, https://history.state.gov/historicaldocuments/frus1938v05/d55.

39. United States, Department of State, "Secretary of State to the American Delegation," 72. In contrast, delegates were warned to be wary of engaging in issues of women's rights, indigenous peoples, and labor conditions.

40. The literature on this is vast, but some representative studies include Lopez, *Crafting Mexico*; and Vianna, *Mystery of Samba*. The Andes is treated in Cadena, *Indigenous Mestizos*. The role of folklore in particular is best developed for Latin America by Mendoza, *Creating Our Own*; and Chamosa, *Argentine Folklore Movement*.

41. Pan American Union, *Travel in the Americas, Report of the Activities of the Travel Division*, n.p.

42. Pan American Union, *Folk Songs and Stories of the Americas*. Remarkably, this publication was revised and reissued in multiple editions, with the last in 1971.

43. Pan American Union, *Holidays and Festivals in Central America and Panama*; Pan American Union, *Holidays and Festivals in Mexico*. Short and published in mimeographed form, these offerings were intended to be sold for a reasonable price (five cents) and distributed widely.

44. Heitor Bastos Tigre, "Some Latin American Festivals: Carnival in Brazil," *Bulletin of the Pan American Union* 73 (November 1939): 649.

45. For the debates within these efforts, see Williams, *Culture Wars in Brazil*. Vianna points out in *Mystery of Samba* that these efforts can also be seen much earlier, at least in the 1920s.

46. Francisco J. Hernández, "Some Latin American Festivals: I. Fiestas in Peru," *Bulletin of the Pan American Union* 73 (November 1939): 643.

47. Mendoza, *Creating Our Own*; Cadena, *Indigenous Mestizos*; Poole, *Vision, Race, and Modernity*.

48. Hernández, "Some Latin American Festivals," 646.

49. This was one of the four pieces later compiled into the volume by the Pan American Union, *Some Latin American Festivals and Folk Dances*.

50. "Folk Dances in Mexico," *Bulletin of the Pan American Union* 73 (February 1939): 96.

51. [Luis Alberto Sánchez], "Some Latin American Festivals: Folk Dances of Spanish America," *Bulletin of the Pan American Union* 73 (November 1939): 652–58. Sánchez had been exiled for his connections with APRA, which may have put him in contact with José Uriel García. Chile's Ministry of Education published the *Revista de Educación* under the direction of Tomás Lago.

52. [Sánchez], "Some Latin American Festivals: Folk Dances," 652.

53. More specifically, the institute also sought to elevate the standard of living of Indians within the Americas. The congress had been recommended by the eighth PAU conference, held at Lima in 1938. The institute was based in Mexico and first headed by Mexican anthropologist Manuel Gamio.

54. Cited in Palomino, "Nationalist, Hemispheric, and Global," para. 30.

55. It is also worth noting that in 1942 the PAU published a curious short work on the indigenous populations of Chile as a part of a series for young readers produced with the cooperation of the Office of the Coordinator of Inter-American Affairs. See Lassalle, *The Araucanians*. The series was composed by diverse authors, including one volume by American folklorist and writer J. Frank Dobie. As these volumes did not advocate travel per se, however, I will not treat them here.

56. Charles Seeger (father to the singer and activist Pete Seeger) was the Music Division's first director. The expanded Division of Music and Visual Arts in 1949 began a new series entitled Folk Songs and Dances of the Americas.

57. Pernet, "'For the Genuine Culture of the Americas,'" 136. Pernet's excellent analysis also surveys the timing of the study of folklore in Latin America and shows how it also coincided with parallel trends within the United States, particularly in the realm of music, but also more broadly.

58. The tensions of this are surveyed best in Andrews, *Afro-Latin America*.

59. Pan American Union, *Brazil* (1937), 15.

60. Zolov, "Discovering a Land 'Mysterious and Obvious,'" 235. Saragoza argues in "Selling of Mexico" that this focus on cultural essentialism had shifted by the 1940s.

61. Jolly, *Creating Pátzcuaro, Creating Mexico*; Chamosa, "People as Landscape;" Romo, "Writing Bahian Identity."

62. She is given authorial credit in later editions but not the first. This was not uncommon in PAU publications, which sometimes list an author and other times simply list the institution itself. I have been unable to ascertain biographical information for Hastings but assume from her name and her tone that she is American.

63. Pan American Union, *Holidays and Festivals in Mexico*, n.p.

64. In an intriguing parallel, musicologist Carol Hess makes the argument that promotions of pan-Americanism in musical terms moved in the 1930s from a universalist rhetoric to a nationalist one, shifting the framing of the music of Latin America from the realm of familiar to difference. Hess, *Representing the Good Neighbor*.

65. Inter-American Travel Congress, First (1939), *Final Act*, 3, 9.

66. The idea of a travel congress had first been suggested to the PAU in 1934 by the Argentine Touring Club, but the idea took shape only after growing recognition by the PAU's Travel Division (established in late 1933) that such a gathering was needed. See José Tercero, "First Inter-American Travel Congress," *Bulletin of the Pan American Union* 72 (August 1938): 465. The conference continues to meet, and its most recent meeting was in 2018.

67. Inter-American Travel Congress, First (1939), *Final Act*, 23.

68. Historian Helen Delpar treats the cultural relations between the United States and Mexico more broadly while highlighting the period between 1927 and 1935 as formative. See Delpar, *Enormous Vogue of Things Mexican*.

69. Schwartz, *Flying Down to Rio*, 333. Racism within the white elite certainly played a part in this concern.

70. I base this observation on the final congress proceedings, although Mexican organizers seem to have perhaps had a different perspective. Historian Dina Berger makes mention of a two-week auto caravan through Mexico organized for U.S. journalists and visitors in 1941, intended to precede and draw attention to the second PAU travel conference, held in Mexico City that year. D. Berger, "Goodwill Ambassadors on Holiday," 119–21.

71. Inter-American Travel Congress, First (1939), *Final Act*, 12.

72. Inter-American Travel Congress, First (1939), *Final Act*, 15, 16.

73. The organization of inter-American athletic events also gained passing mention here. Tercero, "First Inter-American Travel Congress," 474.

74. The *Bulletin of the Pan American Union*, it should be noted, reprinted the Blas illustration from a Latin American publication, *El Boletín Latinoamericano de Música*, initiated by German musicologist Francisco Curt Lange in Montevideo in 1936. The journal published early research on various forms of Latin American music, including many works on folk music and dance. For a useful summary of its early publications, see Berrocal et al., "*Boletín Latino-Americano de Música*."

75. Pan American Union, *Report on the Activities of the Pan American Union, 1938–1948*, 74.

76. In 1947 a reorganization of the Pan American Union that sought to centralize administration divided up responsibilities for travel between the new Department of Economic and Social Affairs (for the elements concerning transportation infrastructure) and the new Department of Cultural Affairs. The next year took new direction as well, with the PAU's reconfiguration as the OAS.

77. M. Camarina de Silva, letter to the editor, *Américas* 1, no. 3 (1949): 48.

78. Inter-American Travel Congress, Third (1949), *Final Act*, 32.

79. Inter-American Travel Congress, Fourth (1952), *Final Act*, 48.

80. Inter-American Travel Congress, Eighth (1962), *Final Act*, 19, 45–46; Inter-American Travel Congress, Ninth (1965), *Final Act*, 39.

81. This was the stated objective of the series; see the inside cover of Pan American Union, *Visit Cuba*.

82. Pan American Union, *Visit Cuba*, 1, 30.

83. Here I am not blind to the way in which folk culture also included elements of exotic display, but I do note a different tone for this later period, one that deserves further study and perhaps represents its own distinct stage of promotion.

84. This section appears to be standardized across the guides. For examples, see Pan American Union, *Colombia*, 31; and Pan American Union, *Brazil* (1965), 29–30.

Bibliography

Aguilar, Alonso. *Pan-Americanism from Monroe to the Present: A View from the Other Side*. Translated by Asa Zatz. New York: Monthly Review Press, 1968.

Albes, Edward. "The Pan American Union." *Southwestern Political Science Quarterly* 1, no. 3 (1920): 248–57.

———. *Rio de Janeiro: The Fair Capital of Brazil*. Washington DC: Pan American Union, 1917.

Andrews, George Reid. *Afro-Latin America, 1800–2000*. New York: Oxford University Press, 2004.

Barrett, John. *Our Opportunity in Latin America*. Washington DC: Pan American Union, 1916. An earlier version was published in *Review of Reviews*, October 1914.

Bendix, Regina. *In Search of Authenticity: The Formation of Folklore Studies*. Madison: University of Wisconsin Press, 1997.

Berger, Dina. "Goodwill Ambassadors on Holiday: Tourism, Diplomacy, and Mexico-U.S. Relations." In *Holiday in Mexico: Critical Reflections on Tourism and Tourist Encounters*, edited by Dina Berger and Andrew Grant Wood, 107–29. Durham NC: Duke University Press, 2010.

Berger, Dina, and Andrew Grant Wood, eds. *Holiday in Mexico: Critical Reflections on Tourism and Tourist Encounters*. Durham NC: Duke University Press, 2010.

Berger, Mark T. *Under Northern Eyes: Latin American Studies and US Hegemony in the Americas, 1898–1990*. Bloomington: Indiana University Press, 1995.

Berrocal, Esperanza, et al. "*Boletín Latino-Americano de Música*." Le Répertoire international de la presse musicale (RIPM), 2014. https://ripm.org/index.php?page=JournalInfo&ABB=BLA.

Cadena, Marisol de la. *Indigenous Mestizos: The Politics of Race and Culture in Cuzco, Peru, 1919–1991*. Durham NC: Duke University Press, 2000.

Cándida Smith, Richard. *Improvised Continent: Pan-Americanism and Cultural Exchange*. Philadelphia: University of Pennsylvania Press, 2017.

Chamosa, Oscar. *The Argentine Folklore Movement: Sugar Elites, Criollo Workers, and the Politics of Cultural Nationalism, 1900–1955*. Tucson: University of Arizona Press, 2010.

———. "People as Landscape: The Representation of the *Criollo* Interior in Early Tourist Literature in Argentina, 1920–30." In *Rethinking Race in Modern Argentina*, edited by Paulina L. Alberto and Eduardo Elena, 53–72. New York: Cambridge University Press, 2016.

Delpar, Helen. *The Enormous Vogue of Things Mexican: Cultural Relations between the United States and Mexico, 1920–1935*. Tuscaloosa: University of Alabama Press, 1992.

———. *Looking South: The Evolution of Latin Americanist Scholarship in the United States, 1850–1975*. Tuscaloosa: University of Alabama Press, 2008.

Drinot, Paulo, and Alan Knight, eds. *The Great Depression in Latin America*. Durham NC: Duke University Press, 2014.

Espinosa, José Manuel. *Inter-American Beginnings of U.S. Cultural Diplomacy 1936–1948*. Washington DC: Bureau of Educational and Cultural Affairs, U.S. Department of State, 1976.

Fagg, John Edwin. *Pan Americanism*. Malabar FL: R. E. Krieger, 1982.

Fifer, J. Valerie. *United States Perceptions of Latin America, 1850–1930: A "New West" South of the Capricorn?* Manchester, England: Manchester University Press, 1991.

Fox, Claire F. *Making Art Panamerican: Cultural Policy and the Cold War*. Minneapolis: University of Minnesota Press, 2013.

Gilderhus, Mark T. *Pan American Visions: Woodrow Wilson in the Western Hemisphere, 1913–1921*. Tucson: University of Arizona Press, 1986.

Guy, Donna J. "The Politics of Pan-American Cooperation: Maternalist Feminism and the Child Rights Movement, 1913–1960." *Gender and History* 10, no. 3 (1998): 449–69.

Hess, Carol A. *Representing the Good Neighbor: Music, Difference, and the Pan American Dream*. New York: Oxford University Press, 2013.

Inman, Samuel. *Inter-American Conferences 1826–1954: History and Problems*. Washington: University Press of Washington DC, 1965.

Inter-American Travel Congress, First (1939, San Francisco CA). *Final Act*. Washington DC: Pan American Union, 1939.

———, Third (1949, San Carlos de Bariloche, Argentina). *Final Act*. Washington DC: Pan American Union, 1949.

———, Fourth (1952, Lima, Peru). *Final Act*. Washington DC: Pan American Union, 1952.

———, Eighth (1962, Guadalajara, Mexico). *Final Act*. Washington DC: Pan American Union, 1962.

———, Ninth (1965, Bogotá, Colombia). *Final Act*. Washington DC: Pan American Union, 1965.

Jolly, Jennifer. *Creating Pátzcuaro, Creating Mexico: Art, Tourism, and Nation Building under Lázaro Cárdenas*. Austin: University of Texas Press, 2018.

Lassalle, Edmundo. *The Araucanians*. Washington DC: Pan American Union, with the cooperation of the Office of the Coordinator of Inter-American Affairs, 1942.

Leonard, Thomas M. "The New Pan Americanism in U.S.-Central American Relations, 1933–1954." In *Beyond the Ideal: Pan Americanism in Inter-American Affairs*, edited by David Sheinin, 95–114. Westport CT: Greenwood Press, 2000.

Lopez, Rick A. *Crafting Mexico: Intellectuals, Artisans, and the State after the Revolution*. Durham NC: Duke University Press, 2010.

Mendoza, Zoila S. *Creating Our Own: Folklore, Performance, and Identity in Cuzco, Peru*. Durham NC: Duke University Press, 2008.

———. "Tourism, Folklore and the Emergence of Regional and National Identities" In *Cultural Tourism in Latin America: The Politics of Space and Imagery*, edited by Michiel Baud and Annelou Ypeij, 23–44. Leiden, Netherlands: Brill, 2009.

Merrill, Dennis. *Negotiating Paradise: U.S. Tourism and Empire in Twentieth-Century Latin America*. Chapel Hill: University of North Carolina Press, 2009.

Palomino, Pablo. "Nationalist, Hemispheric, and Global: 'Latin American Music' and the Music Division of the Pan American Union, 1939–1947." *Nuevo Mundo Mundos Nuevos* (online) June 11, 2015. https://doi.org/10.4000/nuevomundo.68062.

Pan American Union. *Brazil*. Washington DC: Pan American Union, 1937.

———. *Brazil*. Washington DC: Pan American Union, 1965.

———. *Colombia*. Washington DC: Pan American Union, 1966.

———. *Cuba: General Descriptive Data Prepared in June 1911*. Washington DC: Pan American Union, 1911.

———. *Documentary Material for the Good Neighbor Tour: An Imaginary Visit to the Republics of Latin America*. Washington DC: Pan American Union, 1943.

———. *Folk Songs and Stories of the Americas*. Washington DC: Pan American Union, 1937.

———. *Holidays and Festivals in Central America and Panama*. Washington DC: Pan American Union, Travel Division, 1940.

———. *Holidays and Festivals in Mexico*. Washington DC: Pan American Union, Travel Division, 1943.

———. *Inter-American Travel Statistics: Seasonal Movement of Travel in the Americas, January–December 1939, January–June 1940*. Washington DC: Pan American Union, Travel Division, 1940.

———. *List of Publications Published and Distributed by the Pan American Union*. Washington DC: Pan American Union, 1914.

———. *Report on the Activities of the Pan American Union, 1938–1948, Submitted by the Director General to the Member Governments in Accordance with the Resolution of the Fifth International Conference of American States, from the Ninth International Conference of American States, Bogotá, Colombia*. Washington DC: Pan American Union, 1948.

———. *Seeing South America*. Washington DC: Pan American Union, 1927.

———. *Some Latin American Festivals and Folk Dances*. Washington DC: Pan American Union, 1939.

———. *Travel in the Americas: Report of the Activities of the Travel Division of the Pan American Union for the Period January 1–December 31, 1935.* Edited by the Travel Division Pan American Union. Washington DC: Columbus Memorial Library, Organization of American States, 1935.

———. *Viajando por los Estados Unidos.* Washington DC: Pan American Union, 1925.

———. *Visit Cuba.* Washington DC: Pan American Union, Travel Division, 1957.

Pernet, Corinne. "'For the Genuine Culture of the Americas': Musical Folklore, Popular Arts, and the Cultural Politics of Pan Americanism, 1933–1950." In *Decentering America*, edited by Jessica C. E. Gienow-Hecht, 132–70. New York: Berghahn Books, 2007.

Poole, Deborah. *Vision, Race, and Modernity: A Visual Economy of the Andean Image World.* Princeton NJ: Princeton University Press, 1997.

Reid, William A. *Glances at Ports and Harbors around South America.* Washington DC: Pan American Union, 1921.

———. *Mexico, the City of Palaces.* Washington DC: Pan American Union, 1918.

———. *São Paulo, the Heart of Coffee Land.* Washington DC: Pan American Union, 1924.

Romo, Anadelia A. "Writing Bahian Identity: Crafting New Narratives of Blackness in Salvador, Brazil, 1940s–1950s." *Journal of Latin American Studies* 50, no. 4 (2018): 805–32.

Salvatore, Ricardo Donato. *Disciplinary Conquest: U.S. Scholars in South America, 1900–1945.* Durham NC: Duke University Press, 2016.

Saragoza, Alex. "The Selling of Mexico: Tourism and the State, 1929–1952." In *Fragments of a Golden Age: The Politics of Culture in Mexico since 1940*, edited by G. M. Joseph, Anne Rubenstein, and Eric Zolov, 91–115. Durham NC: Duke University Press, 2001.

Schwartz, Rosalie. *Flying Down to Rio: Hollywood, Tourists, and Yankee Clippers.* College Station: Texas A&M University Press, 2004.

Smith, Joseph. "The First Conference of American States (1889–1890) and the Early Pan American Policy of the United States." In *Beyond the Ideal: Pan Americanism in Inter-American Affairs*, edited by David Sheinin, 19–32. Westport CT: Greenwood Press, 2000.

Tota, Antonio Pedro. *The Seduction of Brazil: The Americanization of Brazil during World War II.* Austin: University of Texas Press, 2009.

Vianna, Hermano. *The Mystery of Samba: Popular Music and National Identity in Brazil.* Translated by John Charles Chasteen. Chapel Hill: University of North Carolina Press, 1999.

Williams, Daryle. *Culture Wars in Brazil: The First Vargas Regime, 1930–1945.* Durham NC: Duke University Press, 2001.

Zolov, Eric. "Discovering a Land 'Mysterious and Obvious': The Renarrativizing of Postrevolutionary Mexico." In *Fragments of a Golden Age: The Politics of Culture in Mexico since 1940*, edited by G. M. Joseph, Anne Rubenstein, and Eric Zolov, 234–72. Durham NC: Duke University Press, 2001.

TWO
Developing National Tourism

5

The Making of an Elite Tourist Enclave

Viña del Mar's Miramar Beach, Chile (1872-1910)

RODRIGO BOOTH

Swimming and beach vacations are common leisure activities for millions of Chileans. Despite the fact that they are common pastimes for a large part of society, there is little domestic research into the history of tourism as a subject of study relevant to understanding modern culture. When contemplating the academic prejudices of historiography and the social sciences, we shouldn't be surprised that more value is placed on labor time than leisure time, on the production of industrial goods rather than the consumption of tourism services, on the port rather than the swimming pool, on the house in the city rather than the vacation home. This has led to a paradoxical situation in which tourism studies are considered irrelevant in a country that boasts of the beauty and diversity of its natural landscapes. In an attempt to remedy this omission, this chapter aims to reflect on the construction of the premier modern leisure space: the beach.

According to the French historian Alain Corbin, the beach is a place that has had a variety of meanings for its visitors over the course of history. For Corbin the transformation of this space into a new landscape of touristic consumption can be explained as a consequence of eighteenth-century medical arguments that the cold saltwater of the sea had therapeutic benefits. This, along with the romantic experience of contemplating the landscape and deriving regenerative psychic effects from doing so, transformed the negative image of the sea. Up to that time the sea evoked mythic and real representations such as giant eels, shipwrecks, pirates, storms, and the punitive lesson of the biblical flood.[1] The old fear and rejection of the unknown depths of the ocean were transformed by a complex mutation of perception that drove men and women to utilize the coasts for rest and recreation.

In the last third of the nineteenth century, nearly one hundred years after the beach's entrance onto the European social scene, there was a modification of the perception of South America's coasts. After a timid approach motivated cold seawater therapies, the beaches of the Southern Cone became perennial summer destinations for elites. From that moment on, tourists' colonization of South America's coasts began altering local cultures. For example, the urbanization of Brazil's beaches in the first half of the twentieth century showed for the first time in the continent's history governmental concern for the creation of a mechanism for promoting tourist-oriented businesses and therefore for satisfying the growing demand for leisure.[2] Likewise, nearly halfway through the century the joint efforts of Argentina's public sector and real estate speculators established Mar del Plata as the country's premier city for tourism.[3]

In Chile there has been little research into the role of leisure and tourism in the production of modern culture.[4] This chapter aims to remedy this oversight, putting forward the hypothesis that the social valorization of the pleasure trip is a significant innovation imposed by modernity. As a result of the transformations of summertime social life seen in the last quarter of the nineteenth century, the urban beach was constructed as a tourist enclave, guaranteeing the leisure class's desire for exclusivity. Bringing together many pleasurable summertime activities, the old Miramar beach in Viña del Mar was the country's first space dedicated to the organized production and consumption of tourism by the elite. A review of the history of the most popular beach at the turn of the twentieth century will allow us to understand the social motivations that determined access to leisure among the first Chilean tourists.

Shivers, Cramps, and Drownings: Chilly Early Approaches to the Chilean Coast

The first swimming facilities in Chile did not have the characteristics needed to properly satisfy the needs of tourists. For a large part of the nineteenth century the main coastal tourist destination was the industrialized port city of Valparaíso. Vacationers who traveled there to enjoy the seaside had to deal with an environment polluted by the presence of ships and other economically productive activities. Despite the fact that the mixed-use nature of the harbor impeded swimming, it was nevertheless the primary recreational space on the Chilean coast.

Disregarding the wishes of nineteenth-century swimmers, the government decided to promote the city's industrial infrastructure at the expense of any recreational activities. Between 1868 and 1876 a railway was built along the entire waterfront, which meant exiling recreational activities to the urban periphery.[5] Furthermore, besides the government's lack of concern for tourists, the first ones who came to Valparaíso-area beaches also had to deal with the inconveniences imposed by nature: primarily, the cold seawater.

In contrast to the imaginary hedonistic characteristics of the Chilean sea, the low temperature of the Humboldt current made swimming an often traumatic experience. Rather than the image of pleasure that was promoted, the most common experience of local bathers was that of shivering. The risk of hypothermia generally prevented contact with the water for more than a few minutes, and with few exceptions the rugged topography of the coastline made swimming and diving dangerous. Even the presence of lifeguards did not fully mitigate the risks.

Over time the influence of those who dared to swim in the sea led to representations of immersion in seawater as an unpleasant experience that shocked one's senses. Shivering, as depicted in early twentieth-century fashion magazine illustrations, revealed an ambiguous relationship with beach vacations. If the low temperature of seawater had determined early therapeutic approaches, the promotion of coasts as tourist attractions required swimmers

14. *February (Refusing to go in the water).* From the Cuban weekly magazine Zig-Zag, 1908. Rodrigo Booth collection.

to overlook the shock of the bone-chilling water. The early promotion of the coast as a destination for travelers from inland areas of Chile had the support of doctors who suggested that ablutions of cold water could help cure rheumatic diseases, consumption, and rickets, among other ailments. J. A. García Quintana described the panacea of "hygienism" and its cold seawater therapies in a pamphlet, *Guía de baños de mar y preceptos hijiénicos para las familias i paseantes*. The memoirs of Ramón Subercaseaux, for example, emphasized how unpleasant it was to swim in the sea. In the 1880s he traveled to Valparaíso every summer with his family for a vacation that included a perennial struggle with his mother, who forced him to "enjoy" the ocean. According to Subercaseaux, "The water was extremely cold . . . they had to almost force me in."[6] The adjective "polar," which Eduardo Balmaceda used to describe the temperature of the water at Miramar Beach in the early twentieth century, was symptomatic of the suffering experienced each time he dared challenge the waves of Viña del Mar.[7]

The frequent marine accidents involving people of all social classes also darkened the pages of Santiago's newspapers at the turn of the twentieth century. The cramps caused by the cold water combined with the low skill level of Chile's early recreational swimmers led to a large number of summertime drownings. According to one estimate, in 1909 only 25 percent of men knew how to swim. The evidence available allows us to suppose that the percentage of females able to swim was even lower.[8]

Given the dangerous conditions of prominent coastal recreation centers in the early twentieth century, a profession that sought to save the lives of adventurous vacationers was seen by tourism entrepreneurs as the only way to keep their customers safe. In 1912 Miguel Pérez, a "savior by profession," became one of the country's few lifeguards. His skill allowed him to save the lives of at least fifteen people in the years he worked at El Recreo Beach in Viña del Mar.[9] A beach accident in which a bystander successfully intervened could easily become national news: one thirteen-year-old boy, for example, was congratulated by the president himself after he saved the life of a drowning friend.[10]

15. Lifeguard Miguel Pérez, identified as "a savior by profession," pictured in the publication *Sucesos*, 1916. Rodrigo Booth collection.

But attempts to promote the pleasures of swimming did not include the incorporation of swimming classes into the public school curriculum. This gap in education ensured that swimming would remain an unsafe activity until the end of the 1920s. Deaths by drowning were common every summer.

Polluted water, the government's lack of interest, the cold water, and the rugged coastline make it difficult to explain Chileans' growing affinity for swimming and beach vacations. If the waters of the Pacific along the central coast didn't provide enjoyable conditions for swimming, why did beaches become such fashionable summertime destinations? To answer this question, it is necessary to understand that summertime vacationers were looking for social valorization of leisure and an opportunity to flaunt it. These attitudes seemed to constitute their primary motivations.

The Valorization of Leisure and the Conspicuous Consumption of Pleasure

Seeking to project an image that would imply membership in high society, members of Chile's ruling class showed a persistent interest in expressing elitist pretensions toward the end of the nineteenth century. They thus promoted the construction of spaces whose primary use was the flaunting of material wealth and leisure time. Beaches efficiently met this goal.

To properly reflect on the social image constructed by visits to the beach, it is worth revisiting the arguments put forward by Thorstein Veblen at the end of the nineteenth century. In his classic essay *The Theory of the Leisure Class*, this predecessor of modern sociology identified the bourgeoisie as the most important social group of modernity.[11] In doing so, he emphasized the role of actions by individuals, which had been overlooked by the studies published up to that time. Veblen viewed not only religion, work, and politics as essential concerns of individuals but also sports and other high-profile exhibitions of individual behavior, and he focused in particular on the ritual of social duties.[12]

Even though wealth was the key element in constructing the identity of the nineteenth-century upper class, Veblen argued that their economic capacity was not enough to distinguish themselves from the working class and the rising professional middle class. The activities of those who belonged to this higher social stratum therefore relied on what he termed "conspicuous consumption."[13] The ostentation of spare time and habits associated with rest, as well as the "conspicuous waste" of money, particularly through speculation, led to a profound mental transformation of the commercial bourgeoisie, which from the nineteenth century on contained gentlemen of leisure. The refinement of consumption would lead many wealthy members of the bourgeoisie to obtain luxury goods that had no purpose other than ostentation, thus transforming the images expressed by the era's elites.[14]

A close reading of *The Theory of the Leisure Class* allows us to determine two situations in which elites conspicuously expressed their condition. The first was in the intimacy of private life, with the consumption of luxury goods and other ostentatious products. For Veblen, home furnishings made of exotic wood, pets such as lapdogs, cats, and horses, and even speaking dead languages were some of the luxuries consumed by the leisure class. On the other hand, the need for recognition that would assimilate new habits into those of older ruling classes drove the leisure class to exhibit its wealth through the externalization of pleasure in social life. This group would employ a sophisticated, exclusive sociability, creating semipublic spaces that would allow them to see and be seen by their peers, giving rise to competition for distinction. The materials with which the clothing of the elite was made, the design of its mansions and gardens, the red carpet, and sports such as tennis or golf were but some elements used both in the United States and the rest of Western society to this end.

Veblen formulated his theses by observing the society of his time. His role as a critical witness to the world allowed him to develop categories that could be applied to the Latin American context. In Chile, Luis Barros and Ximena Vergara have argued that the turn-of-the-century realist novel bore witness to leisure's key role in constructing the identity of the oligarchy as a class.[15] The pompous exhibition of elite social habits, the denigration of labor as a means to access luxury goods, and the centrality given to luck or good fortune characterized the dividing line between the leisure class and the working class.[16] Nevertheless, researchers did not study the "valorization of leisure" as an essential element of those spaces that helped construct the oligarchy's identity. The turn-of-the-century tourist would simply be an anecdote in the study of the social life of Chile's Belle Époque.[17]

Perhaps the first intellectual who would connect the rise of the Latin American leisure class with new tourism habits was Jorge Luis Borges. In an article titled "Thorstein Veblen: Teoría de la Clase Ociosa" Borges wrote that in order to maintain their high-class reputation, Argentinians who couldn't afford a trip to the beach would engage in practices that now seem absurd. According to his own memories, the impoverished bourgeoisie "hid in their houses during the hot months so that people would think they were summering on some ranch or in the city of Montevideo."[18] This voluntary imprisonment was accompanied by the dissemination of the last names of these ambitious families to the society pages of the newspaper, thus advertising their alleged departure. Criticizing this peculiar custom, the author reflected on the social climbing of some of his compatriots, who, despite being penniless, could nevertheless maintain the prestige granted by their seasonally appropriate activities. The ruse mentioned by Borges also occurred in Chile, as can be seen in Eduardo Valenzuela's play *Veraneando en Zapallar*, which debuted in 1915. The play recounted the travails of an elite Santiago family who hid in their house all summer. The goal of the voluntary imprisonment was to avoid gossip that they didn't have enough money to summer at the beach in Zapallar.[19]

In Chile, before laws regarding weekends and vacations were passed, summering on the coast was considered to be one of the most *chic* activities of the "cream of society." A hotel, a beach, or a restaurant could be a powerful attraction, both for elites and those who sought inclusion in this group. Travel and beach recreation on Chile's central coast encapsulated the social imaginary fostered by conspicuous consumption. Aspiring elites' presence on the beach would help prove that they were authentic members of the leisure class.

The Tourist Enclave and the Privatization of Miramar Beach

In 1899 the Subercaseaux Mackenna family, worthy representatives of the Chilean bourgeoisie, began a pleasure trip through Europe. Paris, a favorite destination for the Latin American oligarchy at the turn of the century, served as the center of operations in the Old World for the wealthiest representatives of the local elite. Like a fin de siècle parody of the eighteenth-century Grand Tour, the "cultural journey" that the Subercaseaux Mackennas undertook included a trip to the most important monuments of French republican culture, to public parks, and to the pretentious neoclassical architecture that evoked the splendor of the ancient world.

Taking along their camera, the Subercaseaux Mackennas captured snapshots of their visit to show off to their acquaintances in Chile. The photo album they brought back with them contained hundreds of photos of Europe, which joined countless other images of Chilean social life.[20] Making a casual comparison, their travel album illustrated their lives in Paris and Santiago: its first pages show the children playing on the Champs-Élysées and near the Arc de Triomphe; later on these same children are shown at Cerro Santa Lucía and Alameda de las Delicias.

In France the family had the time and money needed to undertake what Anne Martin-Fugier has termed the "summer emigration," which consisted of the mass transit of Parisians to the northern coast on trips motivated by curiosity and leisure.[21] The Subercaseaux Mackennas remained in France during the summers of 1899 and 1900. In a cultural environment that connected their habits to those of French high society, the family summered on the beach, taking advantage of their generous supply of leisure time. Their trips to the beach town of Trouville-sur-Mer are captured in their photo album, bearing witness to their experiences.

In 1901 the Subercaseaux Mackenna family returned to Chile. Although their spending and the time dedicated to their social lives likely decreased, the habits they had developed in Trouville-sur-Mer were kept up, this time in a rustic environment that lacked material development of domestic leisure spaces. The family summered in two places. One was a rural estate in Pirque, some twenty-five kilometers from Santiago, allowing them to visit their relatives and engage in a sophisticated simulacrum of rural life. While there they would go horseback riding, supervise the harvest, and throw open-air Sunday banquets. The other vacation spot was Viña del Mar, which by that time was both the *barrio alto* of the Valparaíso metropolitan area and a tourist destination for the nation's oligarchs.[22] A rough stand-in for the beaches of Trouville-sur-Mer, Miramar in Viña del Mar nevertheless allowed the vacationing family to engage in the same activities they had during their trip to Europe: walks along the beach, the children's timid attempts at swimming, and the contemplation of the vast seascape. Like many of the other happy vacationers in Viña del Mar, the Subercaseaux Mackennas understood that Miramar could satisfy their desires for an environment that would set them apart from their peers.

For Miramar to represent the most developed project of oligarchic tourism in a Chilean city, it was necessary to transform its natural surroundings. Contemporaneous with the expulsion of recreational activities from the Valparaíso harbor, the first references to neighboring Miramar date to the 1870s. Recaredo Santos Tornero depicted it in 1872 as a rocky headland facing the sea.[23] The beach that would become the primary summer destination for the oligarchic elite lacked the comforts needed for swimmers to satisfy their recreational desires. In 1875, less than one year after the approval of the plans for the construction of Viña del Mar, scientist Luis Pomar described the beach in his travel notes as a place "located close to the train station and at the foot of the Callao fortress, frequented by swimmers and families who spend the summer there."[24] As an empty space with no infrastructure of importance, Miramar remained attractively rustic.

But the rusticity of Miramar's landscape was not its only attraction. Its location bestowed it with an atmosphere that set it apart from the rest of Viña del Mar. Watched over by Cerro del Castillo, with the mouth of the Marga-Marga estuary to the north and a

16. View of Miramar Beach. From *Sucesos* magazine, 1916. Rodrigo Booth collection.

rocky outcropping to the south, the difficulty of accessing the beach protected it from the prying eyes of strangers. For a long time the difficulty of reaching the beach was completely disproportional to the economic importance of its visitors. As recalled by Benjamín Vicuña Mackenna, perhaps the most distinguished vacationer at Viña del Mar during the 1880s, elegant socialites had to "wear out their shoes . . . on the bluffs," crossing train tracks, or climbing the hill.[25] The difficulties imposed by its topography and lack of accessibility contributed to high society's particular preference for this destination.

Vacationers swam on a small beach on which two simple wooden structures were built in 1882: one bunkhouse for men and another for women.[26] Utilizing a predictable advertising strategy, a local newspaper contrasted the roughness of Miramar's infrastructure with the beauty and vastness of its views. The Viña del Mar weekly *El Cochoa* referred to the beach under the hill as a place where "a poet would write his best verses." The construction of the first proper road to the beach was described as "extremely daring, its cuts worthy of admiration," and it was argued that the elegance of

17. Swimmers at Miramar Beach. From *Sucesos* magazine, 1916. Rodrigo Booth collection.

this new route would make it a favorite destination of "Villamarinos" on summer afternoons.[27]

Ever since its founding in 1874, Viña del Mar has been home to countless economic activities connected to industrial manufacture. The widespread availability of land, its proximity to the sea, the incorporation of new electric technologies (1882), and the incentives granted by the owners of the old hacienda for the construction of factories gave the city a complex social fabric containing both individuals from Valparaíso high society and the industrial workers employed by their companies.[28] The widespread presence of factories in early Viña del Mar led to the initial transformation of Miramar's rustic natural landscape. In 1883, just south of the beach, Lever & Murphy's Sociedad Maestranza y Galvanización built a factory. The noisy presence of a factory dedicated to the manufacture of metal structures, ships, and automobiles, as well as the presence of its workers, may have led to a loss in the intimacy that the elites required for their social life on the beach.

On the other hand, there is abundant evidence that around the Miramar railway station, some three hundred meters from the beach, there was a working-class neighborhood—clearly "dangerous" in the eyes of the well-to-do—in the last two decades of the twentieth century. The overcrowded, unsanitary slums on the other side of Cerro del Castillo sharply contrasted with the experiences of the Miramar swimmers. The excessive consumption of alcohol made those neighborhoods a permanent scene of quarrels and brawls, not to mention several infamous murders; the Miramar neighborhood beyond the hill was arguably the most dangerous in all Viña del Mar. The information on crime in Miramar was published in *El Comercio* at the beginning of the 1900–1901 high season. This newspaper's primary function was to denounce the lack of safety perceived by those living near Miramar and by visitors arriving by train.

The problems caused by the beach's proximity to industrial activity and crime that exceeded the Miramar police officers' control seem to have been ignored by the beachgoers: the maintenance of a strict security perimeter prevented intruders from taking advantage of the summertime pleasures of Miramar. The tourism promoter Teodoro von Schroeders, in an ever-controversial strategy, ensured that the beach would remain a private, closely guarded tourist enclave, one characterized by fraternization of the richest travelers. The demarcation of Miramar's social environment allowed for a beach that was explicitly offered up for the consumption of elegant turn-of-the-century tourists, contributing to a representation of Viña del Mar that overlooked any activity other than that of summertime recreation.

In 1886 the municipal government awarded von Schroeders a concession that would allow him to exploit a section of Viña del Mar's beaches. For the first time, a private individual would successfully organize a tourism business. Nevertheless, the ambiguous administration of a space that legally should have been considered federal property would lead to many future conflicts.

The first battle between the concessionaire and the municipal government of Viña del Mar occurred in 1888. Understanding that the concession gave him exclusive rights to Miramar Beach, von Schroeders closed it off to public access, sparking an immediate reaction from the local authorities.[29] While it had been stipulated that an entrance fee would be charged in order to allow the business operator to recoup his investment, the privatization of the access route to the beach prompted criticism from members of the public, who considered it their right to go to the beach.

Even though it hasn't been possible to determine the intentions behind the enclosure of the beach, it's clear that von Schroeders's actions were oriented toward creating a safe, secure space that would be comfortably used only by those who enjoyed the privilege of leisure time. In line with this argument, the hiring of private security guards and the construction of a fence contributed to the creation of a semipublic beach. By applying unprecedented "citizen security" measures and guarding the entrance, the businessman's administration encouraged the elitist pretensions of its visitors. The implicit social segregation and illegal enclosures practiced by von Schroeders remained in place until his concession expired in the early twentieth century.

The concessionaire understood that he had territorial rights over the space in which the beach was located. When requesting zoning permits to build new infrastructure for the 1892–93 vacation season, he made constant references to the beach "I own in Miramar."[30] Due to his firm belief that he was a "proprietor," it should not come as a surprise that in 1894 the municipal government of Viña del Mar received complaints that von Schroeders "has built a wooden fence and placed a zinc gate on the road over the hill that goes to Miramar beach, blocking traffic to this and neighboring beaches."[31] The mayor did not hesitate to respond to this infraction. After ordering von Schroeders to destroy the barrier, the municipality sent the local police to force the concessionaire to reopen public access to the beach.

The restrictions imposed by von Schroeders continued until the term of his contract expired. In 1906, for example, the democratic press of Viña del Mar called for an end to the social discrimination seen in Miramar. Criticizing the preferential treatment provided by the new concessionaire, the newspaper *La Defensa* fought to publicize the fact that beach access was the right of all. As the weekly argued in an editorial, "We feel that we have to bring the uncomfortable and hateful distinction that is made at the beach between aristocratic ladies and women of more modest means to the attention of its proprietor, Mr. Mendelewsky. While the former are served with humiliating obsequiousness, the latter are treated insolently."[32]

The security that Miramar guaranteed for vacationers from Viña del Mar allowed it to be the site of some of the most important social gatherings of the first decade of the twentieth century. In 1908 one of the season's charity events brought distinguished visitors to the beach. The protection they requested against intruders had the result of making Miramar "somewhat enclosed to avoid the entrance of any clandestine visitors to the site, leaving a small entrance for the ladies in charge of the sale of tickets."[33] Without the need to hide the segregation and privatization of a supposedly public space, the magazine *Zig-Zag* legitimized the security and coercion practices utilized at the beach.

Protecting the beach from social mixing and the danger of the nearby industrial, working-class environment, the concessionaires took care to ensure that Miramar would meet the needs of the most elegant Viña del Mar vacationers. Around the turn of the century it was common for the society pages to run photographs from the tourist enclave in Miramar. The cross-class coexistence accepted by the rest of the city was in stark contrast to the images of ladies strolling through the sand, children from well-off families playing on the beach, and "gentlemen of leisure" enjoying a happy social life. There would also be a private security guard, dressed like the British colonial police, who was there to ensure that everything was in order. His presence confirmed the importance of security to the swimmers, guaranteeing that Miramar would bring together all the pretensions of a leisure class that wished to engage in the most sophisticated ritual of personal behavior: that of seeing and being seen.

Toward the Construction of the "City of Leisure": Scenes from the Birth of Tourism in Viña del Mar

Oligarchic consumption of tourism created powerful images for those who came to Viña del Mar each summer. The working-class presence that seemed to contradict the city's elegance was not powerful enough to prevent the arrival of vacationers. Even though there are no reliable statistics on domestic tourism dynamics at the turn of the century, there is an abundance of secondary information contained in memoirs, photographs, newspaper articles, and travel guides. Such sources allow us to confidently state that Viña del Mar was the primary destination for the country's gentlemen of leisure. In line with its status as a space of conspicuous consumption by the elite, Viña del Mar contained many luxurious social establishments. Its sophisticated recreational infrastructure contributed to the construction of an image of Viña del Mar as the only "city of leisure" in Chile.[34]

The social importance that elites gave to traditional pastimes brought to Chile by northern European immigrants meant that Viña del Mar was home to the country's first athletic facilities. Fox hunting and the first soccer, tennis, and polo matches, not to mention the paper chase, equestrianism, and the "athleticization" of the traditional horse races held at Valparaíso's Playa Ancha, made the Valparaíso Sporting Club, particularly its racetrack, the leading leisure sports institution in Viña del Mar.[35] Aside from providing recreation for the ruling class, the gatherings held at the club consolidated the social and national identities of local and foreign elites who had come to live or summer in Viña del Mar.

From 1901 on, belonging to a social club provided a secluded social life for vacationers and wealthy Valparaíso business leaders. Located next to the town square and very close to the train station, the Club de Viña del Mar presented itself as the most exclusive local social center. Hosting the most popular recreational activities at a private location that was accessible to only a handful of members, the Club de Viña del Mar was at the forefront of Viña del Mar society until well into the twentieth century.[36]

But the exclusivity of the Valparaíso Social Club and the Club de Viña del Mar did not allow the elites to meet their need to consume conspicuously. While members were able to engage in activities that were the direct opposite of those of labor and production, this voluntary enclosure did not allow members to flaunt this condition before the rest of society. Viña del Mar therefore also contained spaces in which a life of leisure was exhibited, or flaunted, for others to see. Activities at spaces dedicated to tourism became efficient catalysts of ostentation.

In the last decades of the nineteenth century and the early decades of the twentieth, visitors to the town's most expensive hotels could draw out the differences between the leisure class and those who wished to join it. The Gran Hotel de Viña del Mar, built in 1874 next to the railway station and more than two kilometers from the beach, became the principal haven for the ruling class. While the most well-off vacationers occupied the rooms of the Gran Hotel, the rest had to find lodgings in the town's still-unreliable tourism infrastructure. Endowing this establishment with the power to mark social differences, the local press constantly valorized the summertime consumption that took place at the hotel. The publicity given to the establishment by the Viña del Mar press contributed to the ostentation of those activities required of a gentleman of leisure. Strolls through the gardens, a tennis match, banquets, and conversations with other travelers tended to be the primary subjects correspondents discussed in their society page columns.

In 1882 the newspaper *El Cochoa* reported that Benjamín Vicuña Mackenna had rented a cabin in the park of the Gran Hotel.[37] One decade later the visit of Education Minister Federico Errázuriz, who, according to the weekly newspaper *El Viñamarino*, was looking for "health and rest," demanded the publication of a similar article.[38] This journalistic practice, possibly financed by the hotel's owner, was both homage to the subject of the article as well as to the hotel in which he had decided to stay. In this same sense, the publication of the Gran Hotel's guest list confirmed its position as the most prestigious Chilean tourism establishment at the turn of the century. For those who remained in Santiago, unable to take a pleasure trip, reading this list of names, which was also published in the newspapers of the capital, must have had the effect desired by the leisure class: emulation.

A close reading of these guest lists reveals the type of people who took advantage of the beach town's social scene. Of the travelers staying at the Gran Hotel during the 1890s, more than 70 percent had a connection to Anglo-Saxon Protestant culture.[39] Even

18. Games at the Valparaíso Sporting Club. Photographer unknown. Rodrigo Booth collection.

though the information contained on these guest lists did not represent all of the town's visitors, it would be within reason to argue that the seasonal German and English presence contributed to the idiosyncratic atmosphere of the beach neighborhood of Valparaíso.

Social appearances required an eloquent representation of distinction. The refined tastes that lay behind the decision to visit the town and the valorization of leisure by high society triggered the production of a wide variety of journalistic references to those individuals who had simply expressed an interest in making the trip. It should therefore not come as a surprise that the press published lists of those who had made reservations at the Gran Hotel.[40] The planning of a summer visit at Viña del Mar would guarantee a tourist's social ascent. Where and with whom you summered created expectations so powerful that they led to the creation of a true community of gentlemen, which would give Viña del Mar the image of being a "family" swimming destination during the last decades of the nineteenth century.

The hotel publicized these distinctions only through the publication of its guest list. Given that the intraclass fraternization that took place on hotel premises was not easy to exhibit to the rest of society, the importance that the Gran Hotel held for elites lay almost entirely in their ability to make connections with their peers. The inauguration of another hotel on the beach, however, showed the public at large that ostentation was attracting the elite leisure traveler. The Hotel Miramar, the first tourism establishment that successfully integrated the seaside into its design, satisfied the pretensions of those who were interested in conducting their morning walks on the beach. Its low number of beds during the 1890s gave it exclusivity, setting it aside for those who had the economic wherewithal to pay its high prices. Despite the rusticity of its wooden structure, its excellent location on the beach and its proximity to the elegant Schaub Restaurant made it one of the most prestigious lodgings for visitors to Viña del Mar.

The Miramar tourist enclave had attracted the attention of those who followed the summer pleasures of the oligarchy. Visual depictions of the beach abounded at the turn of the century. There was a constant sound of camera shutters as postcard printers, correspondents from the society pages, and countless visitors snapped photos of the coastal landscape and its social life.

As emphasized by a North American visitor who came to Viña del Mar in the early twentieth century, all the pleasant paths led to Miramar on summer afternoons. It was there that local society could happily socialize "outdoors," within sight of the sea. Elegant carriages, ladies dressed up for the occasion, and young men who had come to flirt gave the rustic environment of Miramar an image of careful pomposity.[41]

Revealing the site's roughness, a young woman who signed her name simply as Bessie sent a postcard to her beloved in 1904, telling him that "I spent the happy days of my childhood on this beach." Two years later, a pair of friends sent a postcard in which they discussed the beach's unreliable hotel facilities. None of the postcards had been sent from Viña del Mar. As can be seen on the return addresses, they were sent from the post offices in Santiago and Antofagasta, respectively. As tools of the turn-of-the-century tour-

19. Arriving at the beach in Viña del Mar. From *Zig-Zag* magazine, 1905.

ism industry, postcards guaranteed the dissemination of Viña del Mar's reputation as the country's only large-scale pleasure destination. Given that these postcards were probably sold in the country's biggest cities, their dissemination of images of Viña del Mar formed part of the early repertoire of pleasurable landscapes in Chile. In all, the essentially personal character of the postcard prevented the mass distribution of imagery that beach concessionaires and municipal authorities wished to project. With all the intimacy required by oligarchic tourism, the limited circulation of postcards meant that their images were circulated only among potential visitors.

In 1906 a lengthy article illustrated with photographs of Miramar reported that the beach "becomes an elegant, cosmopolitan world during the summer" and that the inconveniences imposed by its rugged nature "have been overcome by the hand of man."[42] In referencing the work done by the concessionaires von Schroeders and Mendelewsky around the turn of the twentieth century, the magazine *Zig-Zag* emphasized the comfortable facilities for women and men, the lukewarm saltwater baths, the catwalk that had been built in the women's section, the retail establishments along the waterfront, and the terrace that substituted for the boardwalks common at beach destinations elsewhere in the world. While the facilities were not impressive in and of themselves, the beach's elegant visitors made it into the most important social center in Viña del Mar. Through these magazine articles, those who stayed behind in Santiago could "contemplate, perhaps not nature itself, but a copy of it."[43]

Despite the material deficiencies that characterized Miramar as long as tourism remained an oligarchic privilege, the powerful scenes of its social life helped give Viña del Mar an unshakable reputation as a place for the conspicuous consumption of leisure. At least one important work of literature contributed to this process. In her short story "La maja y el ruiseñor," María Luisa Bombal described a walk along the beach in Miramar:

> At any time of day, in groups or couples, hand in hand or walking side by side, the elegant socialites of Santiago and their distinguished counterparts from Viña del Mar came and went all along the beach, bumping into each other and saying hello, and then bumping into each other again, all of them visibly enjoying the air, the sun ... and that social life that was as exclusive as it was pleasurable. The backdrop: palm trees, Victoria cars, friendly drivers and radiant horses bringing or waiting for the happy beachgoers.[44]

One of the first travel guides to be published in Chile gave a similar description. In 1910 *Baedeker de la República de Chile* noted that the beach brought together "the most select strata of the Chilean aristocracy, both in terms of pedigree and money."[45] Putting forward a discourse foreshadowing modern tourism advertising, the travel guide argued that Viña del Mar "is considered to be the Biarritz of the American Pacific: its fame as a luxury beach has gone around the world, with the number of foreign tourists who come to spend the *saison* in this aristocratic *rendez-vous* rising each year."[46] This publicity promoted Viña del Mar as the place that met the summertime recreational expectations of Chileans, especially the elite.

Conclusion

The first places at which Chileans went swimming in the sea were adjacent to cities. Although easy to access, they were often in places with questionable sanitary conditions. Such was the case with Valparaíso, a city where countless members of the nineteenth-century elite spent their summers. Its simple infrastructure was inviting to anyone who wanted to engage in new seasonal leisure activities.

Although the simple tourism infrastructure of the nineteenth century was able to handle the demand represented by the first tourists from Santiago, difficulties imposed by industrial development and the topographical configuration of Valparaíso complicated access to the beach. The cold water, the pollution caused by the industrial port, and the lack of safety measures on the rustic beach discouraged swimming. Paradoxically, in the last third of the nineteenth century, thousands of Chileans began to recreationally enjoy the cold waters of the central coast. Leaving behind the hygienist pretexts that shaped the first approaches to the sea, elites went to the coast in order to flaunt their economic capacity and their place in a modernizing society.

Due to the government's lack of interest, it was private investors who developed the country's first urban swimming beaches. Appearing as an important pillar of the recently founded city of Viña del Mar, the tourism business made Miramar Beach the most prestigious site for enjoying the seaside and its picturesque views. Both the producer and consumer of oligarchic tourism, the Chilean leisure class made this beach into an isolated, guarded space, an enclave of pleasure that met the elite visitors' desire for exclusivity and conspicuous consumption.

The growing valorization of leisure expressed by the turn-of-the-century ruling class led to the development of several spaces in Viña del Mar that were specifically designed to highlight their own ostentation. Hotels, athletic fields, restaurants, and beach amenities became the premier sites at which one could while away the summer. Notable landscape features, efficient administration, and careful promotion made Miramar the place to see and be seen. Like a profane ritual of connection and distinction, these activities helped consolidate the identity of the Chilean elite and of Viña del Mar as the country's only "city of leisure."

Focusing on summertime vacation traditions that are particularly representative of modern social life, this chapter has reflected on the development of the most prestigious beach in Chile at the turn of the century. By examining summertime and weekend leisure, the rise of sports, the construction of new touristic landscapes, and the interurban spatial mobility centered on Santiago, we can better understand the modernization of Chilean culture and society in the early twentieth century.

Notes

1. Corbin, *El territorio del vacío*, 13–85.
2. For the Brazilian case see, for example, O'Donnell, *A invenção de Copacabana*; and Schossler, "Utopias marítimas no Atlântico Sul."
3. Cacopardo, *Mar del Plata*; Pastoriza, *Las puertas al mar*; Bartolucci, *Mar del Plata*.
4. Some exceptions to this rule are Cortés, *Turismo y arquitectura moderna en Chile*; Booth, "Turismo, panamericanismo e ingeniería civil"; and Booth, "'El paisaje aquí tiene un encanto fresco y poético.'"
5. Booth, "El Estado Ausente."
6. Subercaseaux, *Memorias de ochenta años*, 36.
7. Balmaceda, *Un mundo que se fue. . .* , 137.
8. "Los baños y la natación a través de los tiempos," *Zig-Zag*, no. 230 (1909).
9. "Un salvador en el Recreo," *Sucesos*, no. 703 (1916).
10. "Un héroe de trece años," *Zig-Zag*, no. 153 (1908).
11. Veblen, *Teoría de la clase ociosa*. The role of the bourgeoisie in the creation of modern society and as an agent of revitalization and social change, according to Marxist theory, can be seen in Berman, *Todo lo sólido se desvanece en el aire*; and Berman, *Aventuras marxistas*.
12. Hobson, *Veblen*.
13. Veblen, *Teoría de la clase ociosa*, 43–74.
14. Harris, *Jefes, cabecillas, abusones*, 28–32.
15. Barros and Vergara, *El modo de ser aristocrático*, 41–55.
16. Barros and Vergara, *El modo de ser aristocrático*, 50.
17. Cornejo, "Oligarquía y cambio de siglo," 79–102; Vicuña, *La belle époque chilena*, 262.
18. Borges, "Thorstein Veblen," 75–76.
19. Valenzuela, *Veraneando en Zapallar*. This issue has been addressed in greater detail in Booth, "La autosegregación estival y la construcción de la identidad social."
20. I would like to thank the collector Ignacio Corvalán for his knowledge of the images from the Subercaseaux Mackenna family's photo album.

21. Martin-Furgier, "Los ritos de la vida privada burguesa."
22. Cáceres, Booth, and Sabatini, "La suburbanización de Valparaíso y el origen de Viña del Mar"; Cáceres, Booth, and Sabatini, "Suburbanización y suburbio en Chile."
23. Tornero, *Chile ilustrado*.
24. Pomar, "Reconociendo la parte del litoral de Chile," 614.
25. Vicuña Mackenna, *Crónicas Viñamarinas*, 156.
26. *El Cochoa*, January 15, 1882.
27. *El Cochoa*, January 29, 1882.
28. Urbina, "Chalets y chimeneas."
29. Registro de Documentos Municipales, vol. 7, 47, AHVM.
30. Registro de Documentos Municipales, vol. 16, 291, AHVM.
31. Registro de Documentos Municipales, vol. 13, 288, AHVM.
32. *La Defensa*, February 4, 1906, quoted in Rodríguez, "Nuevos aires para un nuevo espíritu," 153.
33. "Kermesse en Miramar," *Zig-Zag*, no. 157 (1908).
34. Booth, "Viña y el mar."
35. Some pastimes popular in Viña del Mar's immigrant communities, and their relationship with local elites during the last decades of the nineteenth century, have been described in Young, *Reminiscences of My Fifty-Five Years in Chile and Peru*. See also "El Valparaíso Sporting Club y su presidente," in *Nuestra Ciudad*, no. 2 (1930). The regulation of pastimes that led to their "athleticization" has been analyzed in Elias and Dunning, *Deporte y ocio en el proceso de la civilización*, 31–81.
36. Salomó, "El Club de Viña."
37. *El Cochoa*, January 8, 1882.
38. "Honorable huésped a del Mar," *Zig-Zag*, no. 49 (1906); *El Viñamarino*, August 23, 1894.
39. This information has been obtained from the lists of guests at the Gran Hotel de Viña del Mar published by local newspapers during the last decade of the nineteenth century. See *La Estación*, November 21, 1892, December 14, 21, 1893; and *El Comercio*, December 6, 1900.
40. *La Estación*, December 14, 1892.
41. Wright, *Republic of Chile*, 224–26.
42. "Viña del Mar," *Zig-Zag*, no. 49 (1906).
43. "En las playas de Miramar," *Zig-Zag*, no. 159 (1908).
44. Bombal, "La maja y el ruiseñor," 15–35. This short story was written in New York in 1959 and originally published in the magazine *Viña del Mar* the following year. "La maja y el ruiseñor" emphasized the Chilean writer's experience in her hometown during the early part of her life. For more information on this important literary figure, see Gligo, *María Luisa*.
45. Sociedad Editora Internacional, *Manual del viajero*, 262.
46. Sociedad Editora Internacional, *Manual del viajero*, 262.

Bibliography

Archival Sources

Archivo Histórico de Viña del Mar, Viña del Mar, Chile (AHVM).

Published Works

Balmaceda, Eduardo. *Un mundo que se fue.* . . . Santiago: Andrés Bello, 1969.
Barros, Luis, and Ximena Vergara. *El modo de ser aristocrático: El caso de la oligarquía chilena hacia 1900*. Santiago: Ediciones Aconcagua, 1978.
Bartolucci, Mónica, ed. *Mar del Plata: Imágenes urbanas, vida cotidiana y sociedad*. Mar del Plata, Argentina: UNMdP, 2002.
Berman, Marshall. *Aventuras marxistas*. Buenos Aires: Siglo XXI, 2003.
———. *Todo lo sólido se desvanece en el aire: La experiencia de la modernidad*. Buenos Aires: Siglo XXI, 1989.
Bombal, María Luisa. "La maja y el ruiseñor." In *El niño que fue*, 15–35. Santiago: Ediciones Nueva Universidad, 1975.
Booth, Rodrigo. "El Estado Ausente: La paradójica configuración balnearia del Gran Valparaíso (1850–1925)." *Eure: Revista Latinoamericana de Estudios Urbano Regionales* 28, no. 83 (2002): 107–23.
———. "'El paisaje aquí tiene un encanto fresco y poético': Las bellezas del sur de Chile y la construcción de la nación turística." *Revista de Historia Iberoamericana* 3, no. 1 (2010): 10–32. http://revistahistoria.universia.cl/pdfs_revistas/articulo_112_1285888012302.pdf.
———. "La autosegregación estival y la construcción de la identidad social: Zapallar y Rocas de Santo Domingo en el proceso de la modernización del ocio en Chile (1892–1950)." *Trace: Travaux et Recherches dans les Amériques du Centre*, no. 45 (2004): 81–92.
———. "Turismo, panamericanismo e ingeniería civil: La construcción del camino escénico entre Viña del Mar y Concón (1917–1931)." *Historia* 47, no. 2 (2014): 277–311.
———. "Viña y el mar: Ocio y arquitectura en la conformación de la imagen urbana viñamarina." *Archivum*, no. 5 (2003): 121–38.
Borges, Jorge Luis. "Thorstein Veblen: Teoría de la clase ociosa." In *Biblioteca personal*, 75–76. Madrid: Alianza, 1998.
Cáceres, Gonzalo, Rodrigo Booth, and Francisco Sabatini. "La suburbanización de Valparaíso y el origen de Viña del Mar: Entre la villa balnearia y el suburbio de ferrocarril (1870–1910)." In *Las puertas al mar: Consumo, ocio y política en Mar del Plata, Montevideo y Viña del Mar*, edited by Elisa Pastoriza, 33–49. Buenos Aires: Biblos, 2002.
———. "Suburbanización y suburbio en Chile: Una mirada al Gran Valparaíso decimonónico." *Archivum*, no. 4 (2002): 151–64.
Cacopardo, Fernando, ed. *Mar del Plata: Ciudad e historia*. Madrid: Alianza-UNMdP, 1997.

Corbin, Alain. *El territorio del vacío: Occidente y la invención de la playa (1750–1840)*. Barcelona: Mondadori, 1993.

Cornejo, Rodrigo. "Oligarquía y cambio de siglo: Mentalidad, costumbres y vida social en Santiago (1899–1901)." Bachelor's diss., Pontificia Universidad Católica de Chile, 1998.

Cortés, Macarena, ed. *Turismo y arquitectura moderna en Chile: Guías y revistas en la construcción de destinos turísticos (1933–1962)*. Santiago: Ediciones ARQ, 2014.

Elias, Norbert, and Eric Dunning. *Deporte y ocio en el proceso de la civilización*. 1986. México DF: Fondo de Cultura Económica, 1995.

García Quintana, J. A. *Guía de baños de mar y preceptos hijiénicos para las familias i paseantes*. Santiago: Imprenta Prat, 1881.

Gligo, Agata. *María Luisa (Biografía de María Luisa Bombal)*. Santiago: Editorial Sudamericana, 1996.

Harris, Marvin. *Jefes, cabecillas, abusones*. Madrid: Alianza, 1985.

Hobson, John Atkinson. *Veblen*. México DF: Fondo de Cultura Económica, 1978.

Martin-Furgier, Anne. "Los ritos de la vida privada burguesa." In *Historia de la vida privada*. Vol. 4, *De la revolución francesa a la primera guerra mundial*, edited by Philippe Ariès and Georges Duby, 226–31. 1987. Madrid: Taurus, 2001.

O'Donnell, Julia. *A invenção de Copacabana: Culturas urbanas e estilos de vida no Rio de Janeiro*. Río de Janeiro: Zahar, 2013.

Pastoriza, Elisa, ed. *Las puertas al mar: Consumo, ocio y política en Mar del Plata, Montevideo y Viña del Mar*. Buenos Aires: Biblos, 2002.

Pomar, Luis. "Reconociendo la parte del litoral de Chile, comprendido entre Viña del Mar i la caleta de Maitencillo por el transporte nacional Ancud." *Anales de la Universidad de Chile*, no. 48 (1876): 583–629.

Rodríguez, Paula. "Nuevos aires para un nuevo espíritu." Bachelor's diss., Universidad Finis Terrae, 2001.

Salomó, Jorge. "El Club de Viña, una tradición centenaria." *Archivum*, no. 2 (2000): 67–75.

Schossler, Joanna Carolina. "Utopias marítimas no Atlântico Sul: Imaginário e tipologias no litoral do Uruguai e do Rio Grande do Sul (1860–1950)." PhD diss., Universidade Estadual de Campinas, 2016.

Sociedad Editora Internacional. *Baedeker de la República de Chile*. Santiago: Imprenta y Litografía América, 1910.

Subercaseaux, Ramón. *Memorias de ochenta años*. Santiago: Nascimento, 1936.

Tornero, Recaredo Santos. *Chile ilustrado: Guía descriptivo del territorio de Chile, de las capitales de provincia, i de los puertos principales*. Valparaíso: Librería i ajencias del Mercurio, 1872.

Urbina, Ximena. "Chalets y chimeneas: Los primeros establecimientos industriales viñamarinos, 1870–1920." *Archivum* 4, no. 5 (2003): 173–96.

Valenzuela, Eduardo. *Veraneando en Zapallar*. Santiago: Imprenta Universitaria, 1915.

Veblen, Thorstein. *Teoría de la clase ociosa*. 1899. México DF: Fondo de Cultura Económica, 1995.
Vicuña, Manuel. *La belle époque chilena: Alta sociedad y mujeres de elite en el cambio de siglo*. Santiago: Editorial Sudamericana, 2001.
Vicuña Mackenna, Benjamín. *Crónicas Viñamarinas*. Valparaíso: Talleres Gráficos Salesianos, 1931.
Young, William Russell. *Reminiscences of My Fifty-Five Years in Chile and Peru*. [Valparaíso]: Sociedad Imprenta y Litografia Universo, 1933.
Wright, Marie Robinson. *The Republic of Chile: The Growth, Resources, and Industrial Conditions of a Great Nation*. Philadelphia: George Barrie and Sons, 1904.

6

"To Know Peru Is to Admire It"

National Tourism Promotion and Populism in Peru, 1930-1948

MARK RICE

In 1941 the Compañía Hotelera del Perú (Hotel Company of Peru) issued a promotional report on its new chain of state-constructed guest accommodations across the nation. The report beckoned to Peruvians to take advantage of the new tourism infrastructure in their country, declaring, "To know Peru is to admire it. Working to know Peru is to teach how to love it; it is to teach to have faith in her grandness; it is to undertake a task of true nationalism."[1] The language of the 1941 hotel report illustrates the degree to which many viewed tourism not only as a new avenue for economic development but as an important tool in forging national identity and politics. Peru's tourism offerings, especially sites like Machu Picchu, are known for their appeal to international travelers.[2] Yet we should not overlook the fascinating history of Peruvian domestic tourism and the role it played in debates over populism, nationalism, and development.

This chapter focuses on the first two decades in which the Peruvian state began to seriously engage with domestic tourism development. In particular this chapter highlights how Peruvian leaders viewed tourism as an important tool for addressing rising populism and social change. Historians of Peru and Latin America have called attention to the ways in which states and political actors responded to the challenge of industrialization and the profound

20. Hotel Company of Peru advertisement for government-built hotels, 1941. From *Cultura Peruana* magazine, December, 1943. Hemeroteca de la Biblioteca Nacional del Perú, Lima, Peru.

social changes it provoked.³ Investigating the role tourism played in these changes connects Peru to the larger global narrative of the 1930s and 1940s. In this era, capitalist, socialist, and fascist states all took an interest in employing tourism to achieve social change and bolster national sentiment.⁴

However, if Peruvian tourism development reflected global patterns, it also revealed the unique political and cultural conditions within Peru as well. As we will see, efforts to develop domestic travel mirrored larger debates over how Peruvian national identity should be imagined. In particular, tourism promotion of Peru's Andean interior sought to elevate a folkloric interpretation of indigenous culture believed to be compatible with a modernizing nation. We will also see that the types of social and racial groups envisioned as the ideal consumers in a new domestic travel economy revealed that tourism backers did not view all Peruvian citizens as eligible to share in this modernization.

Tourism in an Era of Populism

At the start of the twentieth century, a period often called the Aristocratic Republic, Peru's national economy was experiencing modernization and expansion. Yet the leaders of Peru during that era were unwilling to address workers' demands for social and political reform. As a result, a rising tide of populism dominated Peru's national politics during this era. After claiming the presidency in a 1919 coup, Augusto Leguía consolidated dictatorial power, justifying this move as an effort to create a modern state that could manage Peru's economic development and the populist demands of workers. Meanwhile, leftist political leaders like the socialist José Carlos Mariátegui and Victor Raúl Haya de la Torre, founder of the populist Alianza Popular Revolucionaria Americana (APRA) political party, sought to mobilize Peru's citizens in its modernizing coastal cities. Also at this time, in the Andean sierra numerous indigenous agrarian movements were pushing for land reforms and the dismantling of the exploitative hacienda estates.⁵

In the midst of these conflicts the onset of the Great Depression felled Leguía's government and thrust Peru into a period of intense economic and political crisis marked by coups, revolts, strikes, and assassinations. National politics stabilized when Oscar R. Benavides, a former president and army marshal, returned to the presidency in 1933.[6] In many ways the Benavides government represented a return to the politics of the Aristocratic Republic era. He enjoyed the backing of a conservative alliance of economic elites, regional powerbrokers and *hacendados*, and the military. Yet, his allies were not blind to the wave of populism that had begun to wash over Peru. Even before the events of the Great Depression, Peruvian political leaders and intellectuals had realized the need to ameliorate the effects of industrialization through social reforms. As a result, the Benavides government introduced programs of public health, housing, and welfare to benefit Peruvian workers. These reforms were not just attempts at generating goodwill; many of the proposed laws sought to control and discipline workers' populist leanings, and they did little to address the terrible conditions in Peru's Andean agrarian economy. However, the social reforms started under Benavides and extended during the next government, under President Manuel Prado (1939–45), marked a drastic expansion of the Peruvian national state to enact needed social reform.[7]

Tourism, which promised to contribute to the creation of a politically content and physically healthy public through recreation, would naturally appeal to Peruvians of diverse political leanings. For many the promotion of travel promised to help ameliorate the differences between the coast and the sierra. Many Peruvians believed that the differences between the racially mestizo industrializing coast and the largely indigenous agrarian Andean regions had become an unbridgeable social and political divide that prevented the emergence of an effective sense of national identity.[8] Peruvian tourism promoters hoped that, in addition to producing a content and healthy citizenry, travel would also educate citizens and instill a sense of national pride.

One of the most influential voices calling for increased state investment in tourism was the Touring y Automóvil Club del Perú (Touring and Automobile Club of Peru, TACP). Originally formed in 1924 as the Touring Club Peruano to serve as an advocacy group for highway construction and investment, by the 1930s TACP had begun to lobby for the development of a broader tourism economy in Peru. In 1936 TACP transformed its news bulletin into a magazine named *Turismo*; its purpose was not only to provide information on travel in Peru but to shape the nascent tourism industry and state policy toward it.[9] *Turismo*'s first editor was Benjamín Roca Muelle, who had served as minister of finance for part of the Benavides administration and, more importantly, as general manager of TACP between 1939 and 1941. Roca Muelle guided *Turismo*'s editorial profile to echo the aims of TACP in lobbying the state to undertake greater involvement in tourism for the larger economic and social benefit of Peru.

By the mid-1930s Peru was lagging behind its Latin American peers in state efforts to develop tourism. The Cuban government had worked to develop tourism since the first decades of the twentieth century. Localized efforts to develop regional tourism in Mexico also dated to the early 1920s, and that national government had formed a tourism commission by 1928. Both of these countries received economic benefits, an influx of foreign currency, and—at times—even better relations with the United States thanks to growing international tourism.[10] Chile, Argentina, and Uruguay had also committed state resources to tourism by the early 1930s. These countries focused more on developing domestic tourism to spur internal development and to bolster new social welfare programs.[11] In Peru the initial lack of investment in tourism reflected the historical stance of a national elite who since the 1890s had been reluctant to use the state for national development and instead relied on profits from commodity exports and foreign capital investment.[12] This tendency also reflected a national political aversion to social spending and reform on the part of the state as well.[13]

However, with rising political populism and economic distress due to the Great Depression, the ability of tourism to act as a tool of social welfare and as a source of development made it more appealing to national leaders. Responding to increased calls from TACP and a handful of other civic institutions, in 1936 the Benavides government began to examine the possibility of adopting a tourism plan. The Ministry of Foreign Relations commissioned Albert Giesecke, a Philadelphia-born educator who had worked to promote travel to Cusco while he was head of that city's Universidad San Antonio Abad, to travel to the United States in 1936 to study possible strategies for tourism development.[14] Giesecke returned from his mission and gave the Benavides government its desired report. He recommended that Peru dedicate the majority of its efforts to attracting tourism from the United States and argued that the Cusco region should be the primary object of the Peruvian state's tourism plans. This focus on Cusco reflected Giesecke's personal links to the area, as well as the region's early embrace of tourism as a way to promote its cultural folklore and history. While these recommendations would prove to be prescient, Giesecke also acknowledged the importance of the domestic tourism market and suggested that the national state examine the possibility of opening its own hotels in select Peruvian cities.[15]

Encouraged by the findings of Giesecke's report, TACP lobbied the state, especially the Ministry of Development and Public Works, to construct modern hotels in Peru. *Turismo* noted in 1937 that, outside of a handful of hotels in Lima, Peru lacked the type of accommodations modern travelers expected.[16] *Turismo* also noted the direct benefits that a national travel infrastructure promised for Peruvian citizens. The April 1937 edition of *Turismo* featured extensive coverage of tourism policies in neighboring Chile, where state-owned tourist lodges were already operating successfully. Another issue of the magazine included a photo of a new state-built and managed resort with the caption, "Country estate, private property of a major landowner? No, it is the house for all the welfare beneficiaries; far from noise and urban centers, in complete nature, constructed through social security reform to defend health, the precious capital of every worker."[17] The editors of *Turismo* often referenced hotel developments in Chile in the 1930s and 1940s. The February 1946 edition of the magazine published stories on the state-owned hotel at Puyehue, in southern Chile, as well as an extensive report on state-led efforts to promote Viña del Mar as a regional vacation destination. On Viña del Mar *Turismo* noted that "in a friendly and cosmopolitan environment, ladies and gentlemen from around the world gather for celebrations full of happiness and enthusiasm."[18]

Publications like *Turismo* had already printed articles praising Peru's ongoing social welfare reforms, state-led housing construction, and health campaigns. Investment in a domestic tourism program was presented as a continuation of these reforms enacted to create a national workforce that was both physically healthy and politically patriotic.[19] A hotel chain would also provide employment and opportunities. The magazine's editors predicted that modern hotels for tourists would necessitate the hiring and training of new staff, thus creating opportunities for everyday Peruvians—particularly those living outside Lima—to acquire new skills. This training, tourism backers predicted, would provide new educational experiences for hotel and tourism workers, "every one of which," predicted TACP, "is a representation of the culture of the country."[20]

The lobbying of TACP and other groups achieved success. On July 22, 1938, Benavides proposed Law 8708, authorizing the construction of state-owned hotels. Enacted later that year on November 2, the new hotel law authorized the national government to select thirteen sites for new hotels and placed the existing Machu Picchu camping lodge under the ownership of the new organization.[21] In many ways the hotel law marked the first coordinated effort on the part of the Peruvian state to create a national tourism infrastructure. TACP took out a full-page ad in the August 10, 1938, edition of Lima's *El Comercio* newspaper to congratulate the government for addressing the lack of hotels in the interior of the country—a problem the club considered "the biggest obstacle that has existed for the progress of foreign and domestic tourism."[22] Despite this initial focus on attracting foreigners, a heavy emphasis on the importance of tourism as a nationalist project remained constant in press coverage.

By August 1942 the state had completed eight hotels at the cost of roughly 7 million Peruvian soles.[23] The Peruvian state signed an agreement to contract the administration of the hotels to a nonprofit company, Sociedad Hotelera del Perú (Hotel Society of Peru). The membership of the new company was composed of many travel-related institutions, including the owner of the private Hotel Bolívar in Lima, as well as Faucett Aviation, Panagra Airways, and the former head of TACP, Eduardo Dibos Dammert. The Banco Popular del Perú, controlled by the family of President Manuel Prado, also was represented on the board. Significantly, foreign-owned companies that controlled key aspects of Peru's economy, including the Peruvian Corporation and the International Petroleum Company, also had representation on the board.[24] Despite (or perhaps in reaction to) this, hotel proponents emphasized the national character of their development. A report on the hotels printed in *Turismo*'s March 1942 issue reminded readers that "the projects for the construction of these hotels were executed by Peruvian engineers and architects and their construction was overseen by national construction firms."[25]

Getting to Know Peru

For tourism supporters, domestic tourism would not only address populism but serve as part of a larger effort to unify and define the Peruvian nation. The hotel project complemented the Peruvian state's efforts to construct a national road network that backers hoped would form the infrastructural backbone of a unified Peru. As early as November 1937 the Peruvian government, with the support of TACP, launched an advertising campaign, "Conocer y hacer conocer el Perú" (Know and get to know Peru), which encouraged drivers to travel on the new roads and sought to generate public interest in the hotel construction plans.[26] Promotional maps and posters of the hotel construction programs often highlighted each location's proximity to the ongoing road construction efforts.[27] *Turismo*, in a report on the progress of hotel construction, began by listing the driving distance from Lima to the hotels. Some hotel projects were justified as rest stops between cities. Although the report lauded the proximity of the proposed Chala Hotel to the beach, it justified the construction by noting that the site was "located . . . on the trajectory of the Pan-American Highway[;] it is a necessary rest point on the Lima-Arequipa section [of the road]."[28]

Tourism backers hoped that the new road and hotel facilities would encourage Peruvians to visit the interior and gain a better understanding of the country's various regions. Visitors to Peru in the early twentieth century sometimes noted that Lima's elites were more familiar with Paris than the provincial capitals of their own nation.[29] Peruvians were not alone in employing tourism to define regional identity and its place within the nation.[30] However, such a task took on greater importance in Peru, where a lack of defined national identity was often blamed for political and economic instability. Thus, initial promotions of the state-owned hotels also included appealing descriptions of the regions where the new lodgings were placed. A description of the Tumbes Hotel noted the "abundance of fishing" in the coastal region located on Peru's northern border with Ecuador. The description of the hotel planned for the neighboring Piura region reminded readers of the area's "important agriculture and great movement of commerce." Other descriptions highlighted local cultural value and appeal to the nation. Describing the planned hotel for the central-north Andean city of Huaraz, a 1942 report stated that the structure was "located in the Huaylas valley, where one can admire what is, undoubtedly[,] the most beautiful and picturesque [vista] in Peru."[31] Another description of the Huancayo Hotel in the central Andean Mantaro Valley reminded readers that it was located in the "ideal city for tourism, with a dry, healthy, and invigorating climate."[32]

The design of each hotel often sought to give the guest a sense of the character of the region he or she was visiting. The majority of the hotels featured a neocolonial architectural style usually described as *virrenal*. This style, which depicted an idealized version of Peru's colonial past but on a larger scale, already had become emblematic of efforts to remake Lima's historic center in the 1940s.[33] Planners also sought to use each hotel's architecture to reflect the regional features of the area where it was built. A guide noted that the proposed hotel for Arequipa in the south of Peru would be constructed with "a typical *arequipeño* style that has been accentuated with the use of volcanic stones," whose light-gray color would blend in with the exteriors of other buildings of Peru's so-called "White City."[34] An

21. Advertisement for Huancayo, Tingo María, and Huánuco hotels. *Cultural Peruana* magazine, April 1944. Hemeroteca de la Biblioteca Nacional del Perú, Lima, Peru.

article in the Lima magazine *Cultura Peruana* praised the efforts to reflect regional style, noting that the hotel was "harmonizing the sobriety and elegance of its style with the incomparable Arequipa landscape."[35] The hotel for Tingo María, located in the subtropical eastern flank of the Andes, would feature "a rustic style typical of the jungle cabanas situated on stilts." In Ayacucho, a city located in the southern Andes and famous for its historic architecture, the hotel company departed from its usual policy of constructing new buildings and instead renovated the historic Casa de Jáuregui. The project, combining regional history with a new tourist project providing "all of the conditions of comfort and hygiene," illustrated how the goals of nationalism and modernization could be represented by the new hotels.[36]

Promotional materials for individual hotels also sought to emphasize regional identity as part of the travel experience. These images in particular celebrated a folkloric interpretation of each region's indigenous culture. In an era when Peruvians debated how indigenous culture should be represented in national identity, Cusco's elites were some of the first to employ the discourse of tourism to frame a specific interpretation of indigeneity. Increasingly, tourism narratives celebrated a folkloric interpretation of the region's indigenous population as the true representation of Peru.[37] The poster for the Cusco hotel continued this narrative by showing an Inca wall, a llama, and a woman wearing *ñusta* attire, which was patterned on the clothes of female Inca elites and featured prominently in the region's *indigenismo* folklore. Posters for hotel locations show that by the 1940s the Peruvian state had adopted this narrative to celebrate folkloric interpretations of indigenous culture in other regions of Peru. The poster for the Ayacucho hotel featured a woman dressed in traditional Andean textiles and wearing an iconic hat common among many residents who lived in the region. The travel poster for the hotel in the southern Andean region of Puno featured an image of the famous Lake Titicaca as well as the *chullpas* archaeological funerary towers. The poster for the Trujillo installation emphasized the city's rich colonial architecture, as well as artifacts of the Chimú culture of northern Peru.[38]

22. Advertisement for Hotel Cusco. *Cultura Peruana* magazine, July 1944. Hemeroteca de la Biblioteca Nacional del Perú, Lima, Peru.

23. Advertisement for Hotel Puno. *Cultura Peruana* magazine, November 1944. Hemeroteca de la Biblioteca Nacional del Perú, Lima, Peru.

The prominent place of indigenous culture in the hotel promotional materials reflected how tourism policy mirrored national shifts regarding the depiction of indigeneity in Peru. By the 1940s the more radical social demands of indigenismo had receded while regional and national elites supported it as a more moderate cultural, rather than political, movement. Still, the promotion of Andean culture as being distinctly Peruvian represented the gradual embrace of indigenismo by the national state.[39] One of the promised appeals of the Huancayo hotel, according to travel reports, was the ability for travelers to interact with local indigenous residents at the weekly fair, where "Indians[,] festooned with their regional dress, come from faraway highlands carrying their products for sale."[40] The hotel constructed at Puno promised to offer a similar experience for travelers, who could view "the Indians who traffic its streets wearing traditional clothes with characteristic forms and colors, leading their llamas carrying products they will sell at the market."[41]

Travel for Whom?

The descriptions of indigenous culture as an attraction for domestic tourists revealed problematic assumptions regarding which Peruvian citizens would be embarking on journeys. From the start the intended beneficiaries of a modern domestic travel infrastructure and the national narrative it sought to communicate would be upper class, urban, and white. The lack of any state hotels in Lima clearly implied that the expected travelers would be residents of the capital, as well as foreigners. The national state, the Touring and Automobile Club of Peru, and outlets like *Turismo* did not emphasize travel from Peru's interior to Lima. Although advertisements constantly implored readers to "know Peru," they implicitly appealed to Lima residents to know the provinces of their country. No articles appeared in *Turismo* detailing efforts to bring domestic tourists to Lima. So successful were the state hotels in drawing Lima residents into the Peruvian interior that by the mid-1940s private investors had inaugurated similar hotels outside of Lima to appeal to the same market of travelers.[42] Only in October 1944, when the majority of the state-owned hotel chain had been completed, did *Turismo* note the lack of a planned location in the Lima area.[43] Not only were travelers assumed to be from the capital but hotel promotions clearly sought out upper-class consumers. An advertisement promoting the hotel at Nazca pledged that its name, Hotel Company of Peru, "signifies that this establishment is equipped with every modern comfort and the most conscientious service."[44] Advertisements for the hotels always featured photos of well-dressed, usually white, guests.[45]

Efforts to expand tourism development beyond the professional classes of Lima accelerated with the election of José Bustamante y Rivero as president in 1945. Bustamante's successful electoral platform called for populist reforms, greater democratic participation, and a program of state-led economic development.[46] Much to the delight of tourism interests, the new government promised increased state participation in tourism development and an expansion of opportunities for domestic travel. Bustamante's pursuit of these policies mirrored the efforts of other populist governments of Latin America in the 1940s, particularly Cuba and Argentina, in encouraging economic growth and reform through tourism development directed at workers.[47] In November 1945 Bustamante formed a committee to organize a state tourism institution with the cooperation of TACP leadership.[48] At the start of 1946 the national government budgeted 3 million soles for the promotion of tourism, and in April the president submitted a proposal to Congress to create a government-administered tourism office. *Turismo* applauded the effort and called for quick approval in Congress for "the technical organization of tourism that promises to be one of the largest sources of wealth and grandeur."[49] On June 5, 1946, the Congress of Peru passed Law 10556, which created the Corporación Nacional del Turismo (National Tourism Corporation, CNT)—the nation's first state-supported institution tasked with developing a tourism industry.[50] The organization and institutional reach of CNT reflected the wide range of coordination needed for tourism promotion in Peru. The legislation creating CNT gave the new institution advisory duties to review projects with the local archaeology boards (that managed pre-Hispanic sites), historical restoration councils (that managed colonial and postcolonial historical sites), and the Ministry of Development. More important, the CNT assumed administration responsibilities for the Hotel Company of Peru, the distribution of tourism propaganda, the opening of local tourist offices, and the management of a new tourism guide school.[51]

The first general director of the CNT was none other than longtime *Turismo* editor Benjamín Roca Muelle. An enthusiastic advocate for greater government involvement in tourism investment, planning, and promotion, Roca Muelle embraced his new leadership role.[52] In June 1947 CNT organized Peru's first national tourism congress to bring together state and nonstate stakeholders in the fledgling tourism economy. Many of the gathering's activities focused on attracting foreign tourists to bolster economic development.[53] However, the congress also emphasized the need to extend domestic tourism. Speaking on Radio Nacional, Pedro Bentin Mujíca, who was minister of development and chair of the conference, noted that tourism growth promised gains for all Peruvians. "Tourists do not only come to visit our old cities and abundant archaeological remains," he asserted. "The tourist is a man accustomed to comfort and needs good hotels, efficient public services, and modern systems of transport. . . . These services, logically, do not only benefit tourists, but all Peruvians."[54] Roca Muelle also emphasized that, while CNT was created to bring international tourism to Peru, it also would fulfill a domestic travel mission. "A large sector of the country will have the opportunity to become familiar with the national territory and enjoy the enchantment of the voyage," he predicted. Further, Roca Muelle echoed another theme present in earlier hotel promotion by stressing that tourism afforded Peruvians an opportunity to better know their fellow citizens: "Doing this, the state will contribute to a greater physical and spiritual understanding of our people."[55]

Having assumed the presidency, Bustamante led the government's adoption of tourism policies that tourism backers and TACP welcomed. "In each city that has been favored with the construction of a tourist hotel," noted *Turismo* in 1946, "that hotel has come to fill a much felt need for residents and visitors to the point that the majority of the hotels now in service are over capacity."[56] These reports were not exaggerations. Indeed, by 1947 CNT's network of hotels had grown to seventeen locations and the agency had proposed to expand the existing Arequipa location to meet increased demand.[57] By this time Peruvian national efforts to promote domestic tourism had exceeded those of neighboring Chile, which was once looked to as an example to follow.[58]

In addition to increasing the overall tourism infrastructure, CNT worked to expand domestic travel among Peruvians of all classes. In 1946 CNT unveiled a national vacation package that would be administered through participating hotels. Naturally, state-owned hotels were among the lodging accommodations offered. CNT promised to fund 10 percent of the packages, and it offered further subsidies through reduced hotel rates.[59] *Turismo* applauded CNT for "uniting the patriotic efforts that permitted large groups of the middle classes to have greater familiarity with the national territory, while also meeting their desire for relaxation and amusement at a low cost."[60] *Turismo* reported that the first organized outing undertaken as a result of the new vacation plan took place in October 1946. A group of working-class Lima residents were hosted at the privately owned Hostería de Chosica, located just to the east of the capital. The article applauded CNT's work as "one of its most important activities," which would accomplish "the realization of a vast vacation plan destined to favor the employee, giving him the possibility to use the mandatory vacation required by law, to get outside the capital in search of rest for the body and new horizons for better familiarity with the national territory." The article noted that one of the first activities of the tourist group upon arriving at the hotel was "a salute to the flag ... which gave an image of sincere patriotism." The article reported that CNT had already planned future trips to Trujillo, Huancayo, Arequipa, and Cusco. *Turismo* lauded these efforts, arguing that "these trips have the virtue of connecting people, cities, traditions, customs, and folklore to the vast groups of citizens that form our national workforce."[61]

Conclusion

Unfortunately, Peru's experiment in state-led tourism, along with its overall push toward populist government, proved to be short-lived. In October 1948 Bustamante's government fell to a military coup led by General Manuel Odría. The new Odría regime was backed by Peru's economic elite, who aimed to return Peru to an orthodox liberal economy and withdraw from state-led development initiatives like CNT.[62] The Odría regime wasted little time in scaling back CNT's work, as well as the populist initiatives promoted by the Bustamante government.[63] Unlike most other Latin American countries, which continued state investment in tourism in the postwar years, the Peruvian experiment in state-led tourism development proved ephemeral.[64] While the state hotel chain continued its operations, few new locations were opened, as the Peruvian state largely withdrew from tourism promotion for the next decade. When the Peruvian state did engage in tourism development in the postwar era, it did so largely to attract international travelers. The state-owned hotels eventually were sold off by the Alberto Fujimori government in the 1990s. Most of the locations were either were bought by private hotel interests or repurposed for local economic or residential activity.[65]

Although the attempts by the Peruvian state to engage in tourist travel promotion as part of populist policy proved to be brief, much can be learned from that experiment in state-led tourism development between 1930 and 1948. In particular, it shows that the Peruvian state and its leadership were far more attuned to the need for modernization and reform than assessments of their efforts suggest, even if those efforts were cut short. In comparison with many other Latin American countries, twentieth-century Peru had national leaders who often shied away from state-led economic development. Yet, the state's attempts to encourage tourism during the governments of Benavides, Prado, and Bustamante illustrate a growing national consensus about the need to experiment with reform to address the challenges of modernization and populism. The reform aspect of tourism policy was not limited to economic development but also included efforts aimed at forging a Peruvian national identity. In the view of tourism promoters, traveling would permit Peruvians to finally "know Peru," as much of their travel propaganda promised. In an era when indigenismo and regional identity were becoming formalized, travel and tourism could be used to help define and promote these new narratives of national identity.

Tourism policies revealed both the hopes for and flaws in national efforts to reform Peruvian society during the tumultuous years of the Great Depression and World War II. Peru's state-created hotel and travel network, while an impressive accomplishment, was designed to appeal to urban upper and middle classes, especially in Lima. Many of these efforts were earnest attempts to employ the lure of tourism to encourage Lima's residents to become familiar with the regions of Peru and thus establish greater bonds with communities outside the capital. However, the possibility of provincial travel to other regions or to Lima itself was largely dismissed. As Paulo Drinot has argued, reform efforts in Peru often focused on urban and industrialized labor while overlooking rural agricultural work.[66] Tourism policy reflected these same biases. As we can see, tourism policies under three different governments largely dismissed the overwhelmingly indigenous communities of Andean Peru as attractions of, but not participants in, a modern tourism economy. Even when the Bustamante government sought to popularize travel for working-class Peruvians, these efforts were largely focused on urban rather than rural residents. Thus, tourism policies of this era, in seeking to bridge the gap between coastal and Andean Peru, may have inadvertently encouraged its persistence.

Notes

1. *Nuevos Hoteles del Perú para el Turismo* 2, no. 4 (September 1941): 24, Entry 40, Box 657, Ministerio de Fomento y Obras Públicas, RG 229, USNA.

2. Rice, *Making Machu Picchu*.

3. Drinot, *Allure of Labor*; Klubock, *Contested Communities*; Weinstein, *For Social Peace in Brazil*.

4. Baranowski, *Strength through Joy*; Koenker, *Club Red*; Shaffer, *See America First*.

5. Burga and Flores Galindo, *Apogeo y crisis de la república aristocrática*; Klarén, *Peru*, 203–40; Stein, *Populism in Peru*.

6. Drinot and Contreras, "Great Depression in Peru," 102–28; Klarén, *Peru*, 262–76.

7. Drinot, *Allure of Labor*.

8. Cotler, *Clases, estado, y nación en el Perú*; Larson, *Trials of Nation Making*, 195–201.

9. Eduardo Dibos Dammert, "Palabras del Presidente," *Turismo*, July 1936, HBNP. The club changed its name to the Touring y Automóvil Club del Perú in 1939.

10. Schwartz, *Pleasure Island*, 16–38; Berger, "Goodwill Ambassadors on Holiday."

11. Vidal Olivares, "Chile país del turismo"; Piglia, "En torno a los parques nacionales."

12. Thorp and Bertram, *Peru*, 141–44.

13. Burga and Flores Galindo, *Apogeo y crisis de la república aristocrática*; Cotler, *Clases, estado y nación*, 141–83.

14. Ministerio de Relaciones Exteriores del Perú to Giesecke, March 14, 1936, Folio 1, AG-D-71, AHRA.

15. "Informe," July 24, 1936, Folios 26–27, AG-D-071, AHRA.

16. "Editorial," *Turismo*, March 1937, HBNP.

17. "Defensa de la vida, salud, y alegría de los trabajadores es el seguro social," *Turismo*, June 1937, HBNP.

18. "Valiosa influencia ha ejercido el Casino Municipal en el progreso de hermosa ciudad-balneario Viña del Mar," *Turismo*, February 1946, HBNP.

19. "Es preciso meditar serenamente en las ventajas del seguro social," *Turismo*, March 1937, HBNP.

20. "El Touring y Automóvil Club del Perú," *Turismo*, August 1938, HBNP.

21. Ministerio de Fomento y Obras Públicas, *Nuevos Hoteles del Perú para el Turismo* 2, no. 4 (September 1941): 4, Entry 40, Box 657, RG 229, USNA.

22. *El Comercio de Lima*, August 10, 1938, 3, HBNP.

23. "Government Tourist Hotels and the Peruvian Hotel Company," *Turismo*, August 1942, HBNP.

24. "Government Tourist Hotels and the Peruvian Hotel Company."

25. "El plan hotelero del gobierno del Perú," *Turismo*, March 1942, HBNP.

26. "Editorial," *Turismo*, November 1937, HBNP.

27. "Plan del estado para la construcción de hoteles," *Turismo*, March 1938, HBNP.

28. "El plan hotelero del gobierno del Perú."

29. Franck, *Vagabonding down the Andes*, 426.

30. See Covert, *San Miguel de Allende*; and Brown, *Inventing New England*.

31. "El plan hotelero del gobierno del Perú."

32. "El hotel para turistas 'Huancayo,'" *Turismo*, April 1945, HBNP.

33. Ramos, "La reforma neocolonial de la Plaza de Armas," 101–41.
34. "El plan hotelero del gobierno del Perú."
35. "Hotel de turistas Arequipa," *Cultural Peruana*, February 1941, HBNP.
36. "El plan hotelero del gobierno del Perú."
37. Mendoza, *Creating Our Own*, 65–91; Rice, *Making Machu Picchu*, 16–44.
38. "En la Corporación Nacional del Turismo," *Turismo*, September 1946, HBNP.
39. de la Cadena, *Indigenous Mestizos*; Lauer, *Andes imaginarios*; Mendoza, *Creating Our Own*.
40. "El plan hotelero del gobierno del Perú."
41. "El plan hotelero del gobierno del Perú."
42. "La hostería de Moyopampa, en Chosica," *Turismo*, October 1944, HBNP.
43. "Por un hotel en La Punta," *Turismo*, October 1944, HBNP.
44. "Compañía Hotelera del Perú, S.A. advert," *Turismo*, April 1948, HBNP.
45. "Hotel Paracas," *Turismo*, May 1946, HBNP.
46. Klarén, *Peru*, 289.
47. Elena, *Dignifying Argentina*; Skwiot, *Purposes of Paradise*, 146–47.
48. "Notas y Comentarios," *Turismo*, November 1945, HBNP.
49. "Notas y Comentarios," *Turismo*, February 1946, April 1946 (quote), HBNP.
50. "Memoria de la Corporación Nacional de Turismo, correspondiente al ano 1946," 1947, 9, lejado 2444, H-6, FMH, AGN.
51. "El Gobierno reglamenta la Corporación Nacional de Turismo," *Turismo*, June 1946, HBNP.
52. "Benjamín Roca Muelle," *Turismo*, June 1946, HBNP.
53. "El Primer Congreso Nacional de Turismo," *La Prensa* (Lima), June 9, 1947, 5, HBNP.
54. Quoted in "Primer Congreso Nacional de Turismo," *El Comercio de Lima*, June 7, 1947, 3, HBNP.
55. Quoted in "Benjamín Roca Mulle," *Turismo*, June 1946, HBNP.
56. "Actuales establecimientos," *Turismo*, July–August 1946, HBNP.
57. "Aplicación del actual Hotel de Arequipa," *Turismo*, March 1947; "Compañía Hotelera del Perú, S.A." *Turismo*, July 1947, both in HBNP.
58. "Organización Nacional Hotelería, S.A. advert," *Turismo*, October 1948, HBNP. By this time Chile had five state-owned hotels.
59. "Plan vacacional," *Turismo*, October 1946, HBNP.
60. "Notas y comentarios," *Turismo*, September 1946, HBNP.
61. "La hostería de Chosica recibió el primer contingente de turistas," *Turismo*, October 1946, HBNP.
62. Klarén, *Peru*, 289–99; Thorpe and Bertram, *Peru*, 201.
63. "Notas y comentarios," *Turismo*, October 1948, HBNP.
64. Armas Asín, "Lo esperable del estado"; Rice, *Making Machu Picchu*, 70–74.
65. Rice, *Making Machu Picchu*, 148–49.
66. Drinot, *Allure of Labor*.

Bibliography

Archival Sources

Archivo General de la Nación, Lima, Peru (AGN).
    Fondo Ministerio de Hacienda (FMH).
Archivo Histórico Riva Agüero, Lima, Peru (AHRA).
    Colección Albert Giesecke (AG).
Hemeroteca de la Biblioteca Nacional del Perú, Lima, Peru (HBNP).
United States National Archives, College Park, Maryland (USNA).
    Record Group (RG) 229, Records of the Office of Inter-American Affairs.

Published Works

Armas Asín, Fernando. "Lo esperable del estado: Políticas públicas y empresarios en los inicios de la actividad turística del Perú (1930–1950)." *Apuntes: Revista de ciencias sociales* 46, no. 85 (July 2019): 53–78.

Baranowski, Shelley. *Strength through Joy: Consumerism and Mass Tourism in the Third Reich*. Cambridge MA: Cambridge University Press, 2004.

Berger, Dina. "Goodwill Ambassadors on Holiday: Tourism, Diplomacy, and Mexico-US Relations." In *Holiday in Mexico: Critical Reflections on Tourism and Tourist Encounters*, edited by Dina Berger and Andre Grant Wood, 107–29. Durham NC: Duke University Press, 2010.

Brown, Donna. *Inventing New England: Regional Tourism in the Nineteenth Century*. Washington DC: Smithsonian Books, 1997.

Burga, Manuel, and Alberto Flores Galindo. *Apogeo y crisis de la república aristocrática: Oligarquía, aprismo, y comunismo en el Perú, 1895–1932*. Lima: Ediciones Rikchay Perú, 1980.

Cotler, Julio. *Clases, estado, y nación en el Perú*. Lima: Instituto de Estudios Peruanos, 1975.

Covert, Lisa Pinley. *San Miguel de Allende: Mexicans, Foreigners, and the Making of a World Heritage Site*. Lincoln: University of Nebraska Press, 2017.

de la Cadena, Marisol. *Indigenous Mestizos: The Politics of Race and Culture in Cuzco, Peru, 1919–1991*. Durham NC: Duke University Press, 2000.

Drinot, Paulo. *The Allure of Labor: Workers, Race, and the Making of the Peruvian State*. Durham NC: Duke University Press, 2011.

Drinot, Paulo, and Carlos Contreras. "The Great Depression in Peru." In *The Great Depression in Latin America*, edited by Paulo Drinot and Alan Knight, 102–28. Durham NC: Duke University Press, 2014.

Elena, Eduardo. *Dignifying Argentina: Peronism, Citizenship, and Mass Consumption*. Pittsburgh PA: University of Pittsburgh Press, 2011.

Franck, Harry A. *Vagabonding down the Andes: Being the Narrative of a Journey, Chiefly Afoot, from Panama to Buenos Aires*. New York: Century, 1917.

Klarén, Peter. *Peru: Society and Nationhood in the Andes*. New York: Oxford University Press, 1999.

Klubock, Thomas Miller. *Contested Communities: Class, Gender, and Politics in Chile's El Teniente Copper Mine, 1904–1951*. Durham NC: Duke University Press, 1998.

Koenker, Diane P. *Club Red: Vacation Travel and the Soviet Dream*. Ithaca NY: Cornell University Press, 2013.

Larson, Brooke. *Trials of Nation Making: Liberalism, Race, and Ethnicity in the Andes*. Cambridge: Cambridge University Press, 2004.

Lauer, Mirko. *Andes imaginarios: Discursos del indigenismo 2*. Cusco: Centro de Estudios Regionales Andinos, Bartolomé de las Casas, 1997.

Mendoza, Zoila. *Creating Our Own: Folklore, Performance, and Identity in Cuzco, Peru*. Durham NC: Duke University Press, 2008.

Piglia, Melina. "En torno a los Parques Nacionales: Primeras experiencias de una política turística nacional centralizada en la Argentina (1934–1950)." *Pasos: Revista de Turismo y Patrimonio Cultural* 10, no. 1 (2012): 61–73.

Ramos, Horacio. "La reforma neocolonial de la Plaza de Armas: Modernización urbana y patrimonio arquitectónico en Lima, 1901–1952." *Histórica* 40, no. 1 (2016): 101–41.

Rice, Mark. *Making Machu Picchu: The Politics of Tourism in Twentieth-Century Peru*. Chapel Hill: University of North Carolina Press, 2018.

Shaffer, Marguerite S. *See America First: Tourism and National Identity, 1880–1940*. Washington DC: Smithsonian Books, 2001.

Schwartz, Rosalie. *Pleasure Island: Tourism and Temptation in Cuba*. Lincoln: University of Nebraska Press, 1997.

Skwiot, Christine. *The Purposes of Paradise: U.S. Tourism in Cuba and Hawai'i*. Philadelphia: University of Pennsylvania Press, 2012.

Stein, Steve. *Populism in Peru: The Emergence of the Masses and the Politics of Social Control*. Madison: University of Wisconsin Press, 1980.

Thorp, Rosemary, and Geoffrey Bertram. *Peru, 1890–1977: Growth and Policy in an Open Economy*. New York: Columbia University Press, 1978.

Vidal Olivares, Patricia. "Chile país del turismo: El rol del estado y representaciones sobre Chile en el fomento de una industria moderna." PhD diss., Pontificia Universidad Católica de Chile, 2017.

Weinstein, Barbara. *For Social Peace in Brazil: Industrialists and the Remaking of the Working Class in São Paulo, 1920–1964*. Chapel Hill: University of North Carolina Press, 1996.

7

Domestic Tourism in Golden-Age Veracruz, Mexico

ANDREW GRANT WOOD

> Given ample international promotion, Veracruz could turn into a major tourist attraction given the pleasant climate, beaches, warm hospitality and provisioning of local businesses.
>
> —RUBÉN C. NAVARRO, 1946

By the early 1950s tourism in Mexico had apparently come of age, as evidenced by the growing numbers of international travelers seeking relaxation, sport, culture, and diversion at any number of attractive destinations across the country. Mexico had entered a period of unprecedented economic growth, with an expanding middle class and a national entertainment industry that proudly personified the new "golden" era of prosperity. Paparazzi and camera crews captured images of international celebrities, including Johnny Weissmuller, Elizabeth Taylor, Errol Flynn, Rita Hayworth, Cary Grant, and others, vacationing in Acapulco. Popular magazines published gossip about and glimpses of glamorous stars in Mexico City nightclubs such as El Patio, Sans Souci, and Ciro's. Yet, while media attention and government programming tended to identify tourism as an engagement with foreign visitors, the vast majority of those engaging in leisure-time travel and tourism in Mexico were Mexican.

Domestic travel to Veracruz increased significantly after World War II as transportation services and hotel accommodations improved.[1] International service to and from the Veracruz airport began in mid-1949 with flights from the Caribbean and Central America, as well as from the United States and Canada.[2] To be sure, the city of Veracruz soon attracted its share of international visitors, including, for example, an assortment of sporting types arriving for an array of annual fishing tournaments.[3] The excitement of these and visitations by U.S., Cuban, and other foreign delegations notwithstanding, it was domestic tourism that proved most significant in postwar Veracruz.[4]

Of the estimated sixty thousand Mexican excursionists circulating throughout the country prior to Easter, that is, during Holy Week (hereafter, Semana Santa) in April 1952, industry officials calculated that Veracruz alone would play host to nearly half—that is, somewhere between fifteen thousand and twenty-five thousand travelers would come to the city to enjoy the beaches, heritage sites, parks, cinemas, nightclubs and open-air restaurants.[5] Given the city's only partial success in transforming itself into an international center for modern travel, the spring holiday would nevertheless prove highly successful in developing a domestic tourist trade.

Early Tourism Histories

Travel by horse-drawn coach or on muleback notwithstanding, tracing the origins of modern tourism in Veracruz begins with the advent of steamship and rail travel in the nineteenth century. In 1866 negotiations for a sea route between New Orleans and Veracruz took shape and soon led to increased traffic between the two cities (not to mention Havana and other Caribbean destinations). Railway construction did not lag much behind, and by the mid-1870s an initial connection between the port of Veracruz and Mexico City had been realized. The advent of these key infrastructural advances proved foundational, allowing the city and its residents to begin, however modestly, a trade in commercial tourism. Civil war in the second decade of the twentieth century interrupted this

24. Villa del Mar, ca. 1925. Photographer unknown. Andrew Wood collection.

process, but in the 1920s an enterprising new generation of Veracruz developers again took up the torch.

Tourist development in the port of Veracruz required construction of new urban infrastructure: power lines, water mains, sewage systems, rail lines, roads, clean restaurants and hotels, safe beaches, engaging recreational facilities, and ultimately an airport. Such undertakings required significant investment by federal, state, and local governments in cooperation with private business.

At the national level the founding of the Comisión Mixta de Pro-Turismo and the Comité Nacional Pro-Turismo in the late 1920s under the aegis of the Ministry of the Interior (Secretaría de Gobernación) helped animate talk of tourism with a focus on road and hotel construction. Promoters set out to highlight the state's natural beauty and cultural heritage in a series of marketing campaigns designed primarily to attract domestic tourists. A June 1930 bilingual publication issued by *La Revista Nacional de Turismo*, for example, described the port of Veracruz as "having good hotels, sea bathing and other delightful features." Mention was made of the colonial fortress San Juan de Ulúa as an interesting local attraction, while overall Veracruz "was [to be considered] a popular destination for those seeking a weekend excursion from the capital."[6]

Despite promising initiatives and the relative progress achieved during the 1920s, tourism industry growth was slowed by the global economic downturn in the 1930s. Responding to a state government mandate in late December 1930 asking locals to identify and report on areas of possible interest to tourists, towns across the state of Veracruz, including Cotaxtla, La Antigua, Jáltipan, San Salvador, Alvarado, San Cristobal, Tlacotalpan, and Puerto México (Coatzalcoalcos), helped provide useful information for officials intending to help stimulate industry growth. Veracruz officials had issued a subsequent request for information by late May 1931, asking that key architectural and heritage sites (i.e., churches, hospitals, and other interesting structures) be described, along with the condition and availability of nearby roads and transportation routes. Thus, by 1935 a relatively clear map of tourist attractions and basic infrastructure had begun to take shape. In the postwar years there would be renewed attention to developing tourism in Veracruz.

Meanwhile, private industry enthusiasts and a host of other interested observers were promoting Veracruz as an up-and-coming destination for both domestic and foreign travelers.[7] In 1933, for example, diplomat Halleck L. Rose, who worked in the U.S. vice consul office in Veracruz, wrote in the English-language periodical *Real Mexico* that "Veracruz had the possibility to emerge as a major tourist destination." Rose described tourist conditions in and around the port of Veracruz and offered a relatively positive evaluation, noting the city's oceanside location and observing that "swimming is safe," as local waters are "protected from sharks and other [undesirable] sea life." The vice consul highlighted the area's sport fishing, noting that "many excursions are available as well as many different types of fish." Although noting the absence of golf facilities, he thought highly of local tennis facilities and players. Other recreational facilities remarked upon by Rose included the Villa del Mar "sea-side dance hall with live music and robust local attendance." Overall, Rose concluded that "Veracruz is a good place for a vacation and that [tourism] is good economically for Veracruz."[8]

While the U.S. vice consul praised Veracruz as an attractive tourist destination, he nevertheless made note of the city's dismal reputation as an unhealthy place. Rose observed that "there are many tourists who arrive by boat and stay in the port for the shortest time only to take the train to Orizaba and then on to Mexico City. They know the history of poor health in the port and even though the Sanitation Commission has virtually eliminated mosquitos, the bad reputation remains—even though this is one of the few places where you don't have to sleep with mosquito netting."[9] The vice consul's comment touched upon an age-old problem for port residents and visitors alike: sanitation. In the years to come, improving public health conditions, while at the same time working to give the city a reputation as a safe and reasonably hygienic place, would prove to be a significant undertaking.

Postwar Challenges

As government efforts intended to expand key commercial sectors of the national economy during and after World War II took effect, the local economy of Veracruz grew as well.[10] Given prewar efforts to develop local hospitality, public and private concerns sought to remake postwar Veracruz into an attractive destination for travelers. A 1946 article in *El Dictamen* by guest writer Rubén C. Navarro, who was the Mexican consul general to Brazil, touted the port as a burgeoning center of tourism. "Given ample international promotion," Navarro figured, the city of Veracruz "could turn into a major tourist attraction." He noted that the city offered "pleasant climate, beaches, warm hospitality and provisioning of local businesses." What was needed, Navarro acknowledged, were additional hotels, restaurants, and accommodations, along with effective means of promoting the city.[11] Measures aimed at achieving improved sanitation, increased tourism infrastructure, and a robust marketing campaign all were put into motion in the hope of stimulating industry development. Going forward, Veracruz faced many challenges in seeking to refashion itself as a modern tourist destination.

In October 1946 members of the Mexican Hotel Association (Asociación Mexicana de Hoteles) traveled to the port of Veracruz for their annual convention. Among the local lodgings represented

25. Hotel Diligencias and Veracruz city center, ca. 1950.
Photographer unknown. Andrew Wood collection.

were Hotel Mocambo, Mena Brito, Villa del Mar, Royalty, Cristóbal Colón, Villa Rica, Malecón, Victoria, Buena Vista, Oriente, México, Colonial, Imperial Braña, Terminal, Plaza, Principal, Lux, Miramar, Ortiz, and Veracruz.[12] On the opening night of the gathering, Mayor Ulises Rendón, along with Adolfo Ruiz Cortines, who was governor of the state of Veracruz (and president of the republic in late 1952), welcomed the hoteliers. Ensuing presentations and talks during the Veracruz meeting anticipated a growing number of recreational travelers in Mexico for the years to come.[13]

Hotel owners confidently asserted that a wide cross section of Mexican society working in "industry, finance and business along with local labor unionists and the general public" would benefit from the enhancement of postwar tourism. Expansion of the industry, they figured, would prove profitable not only for hotel owners but also a wide range of related enterprises.[14] For those gathered in the port that October, "tourism constituted an industry of the first order" and Veracruz stood as a leading "symbol" for its future growth.[15] Also present at the hotel owners convention was Francisco C. Lona, who spoke to the assembly on behalf of the National

Railways of Mexico. In his address Lona urged his audience to support tourism development in Veracruz: "I came to the port for the first time in 1919 and ever since then I have appreciated the city's beauty. Now in 1946 I am convinced Veracruz has all the right elements to be an attractive tourist destination."[16]

The port of Veracruz had in fact drawn approximately fifty thousand visitors so far that year. Revenue generated by tourism was estimated to be around 8 million pesos (about $2 million in U.S. dollars). This figure of course represented a percentage of Mexico's overall tourist traffic distributed across a variety of key destinations, including Mexico City, Acapulco, Cuernavaca, Taxco, Puebla, and elsewhere.[17]

Despite the fact that much work needed to be done, those associated with the industry nevertheless continued to promote the port. In a statement circulated among local businesses in late January 1947, Alejandro Buelna, head of Mexico's Department of Tourism, indicated that Veracruz possessed many of the features required of a successful tourist site: water sports, fishing, spas, and modern hotels with affordable prices. What was needed, he asserted, was widespread marketing.[18] A few days later *El Dictamen* published a piece on the Candelaria festival in the neighboring town of Tlacotalpan. Author Felipe Lagardy described the annual event as "romantic," even "mystical," and encouraged readers to make the journey by car or bus from Veracruz, Jalapa, Córdoba, Orizaba, and even Puebla.[19] A short time later it was estimated that approximately ten thousand people from across Mexico, as well as from the United States and other nations, would soon visit the port of Veracruz for the yearly pre-Lenten Carnival celebration.[20]

Still, expansion of local industry required additional hotel rooms, enhanced transport, and improved hospitality services. Furthermore, promoters realized that in order to facilitate tourism locally they needed also to promote regional attractions. Enterprising individuals suggested that the port of Veracruz could conveniently serve as a base for individuals wishing to visit the neighboring Sotovento region, an area south of the port that included the towns of Medellín, Alvarado, and Tlacotalpan. Travel to one of these easy-to-reach destinations, it was understood, would provide tourists an attractive glimpse of Veracruz regional (*jarocho*) hospitality, given the area's unique mix of European, African, and American heritage. Confident in the character and charm of these smaller Veracruz settlements, some even referred to the alluring Sotovento at the time as "the Andalucia of Mexico."[21]

The Alemán Years

On December 1, 1946, hotel owners had taken out a full-page ad in *El Dictamen* to congratulate Veracruz-born (and former governor from 1936 to 1940) Miguel Alemán Valdés on his inauguration as Mexico's new president. Over the course of his ensuing six-year term, Alemán would oversee several important domestic tourism initiatives, including major road, hotel, and other infrastructural projects.

In Veracruz that same month word had circulated about proposed improvements to the city's seaside tourist zone (Zona Maritima) undertaken by the firm of Juan B. Hernández and Guillermo Gómez Cepeda Compañía Turistica Veracruzanos S.A. de C.V.[22] Ambitious plans even included possibly developing the tiny offshore Isla de Sacrificios into a beachfront recreational space. Industry enthusiasts envisioned a major tourism project to include the construction of hotels, bungalows, boating and swimming areas, spas, and more. However, while the realization of such goals remained in question, more immediate concerns occupied city and industry planners.

Issues related to sanitation and public health continued to slow development. A mid-May 1947 headline in *El Dictamen* declared, "Before anything else, the city needs to be cleaned up[,] with priority given to water and drainage service."[23] The article went on to explain how Veracruz leaders—including newly elected mayor Tuero Molina, as well as Enrique Viesca, a key member of the Federal Improvement Association (Junta Federal de Mejoras Materials)—had identified the area just outside the city slaughterhouse as an immediate threat to public health.

A subsequent report made clear that city government needed additional revenue if improvements to the appearance and function of city were to be realized. Streets, avenues, neighborhoods, schools, and public security would all require significant investment in improvements, observed Mayor Molina, if Veracruz was to achieve recognition as a viable tourist destination.[24]

Clearly, considerable investment and work would be needed for the city to become safe, clean, and attractive for visitors. Francisco García Marchena, head of the state tourism office in the port of Veracruz, noted recent declines in tourist traffic to the city and listed what he understood to be several key causes for the drop: "Foreign travelers who have the money to visit are often disappointed by what they see as a lack of sanitation, scarce gasoline supply and the possibility of being taken advantage of."[25] Nevertheless, García Marchena noted that the majority of visitors to the port were Mexican nationals taking time to visit family and to vacation during Carnival, Semana Santa, Fiestas Patrias (Independence Day), Christmas, and the New Year holiday.[26] If Veracruz could not rise to prominence as a destination for wealthy international travelers, perhaps the city could still successfully cater to Mexican nationals seeking fun, relaxation, and family time away from home.

Soon thereafter Miguel Ángel Rodríguez, secretary of the Veracruz Hotel Association, was blunt in his call to action. Overall revenue generated from tourism could significantly increase, he argued, if local infrastructure could be enhanced. For the moment, however, Rodríguez estimated that Veracruz did not have sufficient facilities or the necessary attractions. "People in the city need to do more than just create committees and organizations," he charged. "They need to coordinate, plan, invest and build."[27] Focusing specifically on what he saw as a shortage of available accommodations as the principal cause for the slowdown, Hotel Victoria owner Juan M. Plaza believed that "if, together, we could add another 850 rooms [to the city listing] we would then add an estimated 10.5 million pesos to our revenue stream."[28]

Yet the problem lay not only in a general lack of sanitation or with the perceived lack of hotel space. The challenge for tourism boosters was more complicated than that. Visiting the port from Mexico City, Dr. José Palacios Macedo offered his perspective on

Veracruz hospitality: "Having not been in the port for about ten years, I immediately noticed new construction along the Malecón and several new hotels. But despite this, there remained many rundown tenements as well as several noisy, deteriorating trolleys and filthy carts. Added to that is the same crappy service in the restaurants!"[29] Palacios's commentary suggests the city still had a long way to go before it would be the kind of attractive tourist destination some imagined. Loud, dirty, somewhat inhospitable, and lacking in services, the port of Veracruz also had a tuberculosis problem, with nearly three thousand affected.[30]

Campaigns aimed at cleaning hotels, guesthouses, restaurants, and even private homes in popular neighborhoods were undertaken—both to contain the spread of the infectious disease and also in the hope of eventually attracting tourists.[31] For their part, Veracruz renters at the time again organized to address the city's persistent shortage of affordable and sanitary housing.[32] Given these and other endeavors, public health officials made a point of periodically reporting the results of various sanitation projects. One such installment a few years later contained the proud declaration that "12 thousand tons of garbage [had been] removed from the port."[33]

Discussion and action in Veracruz in the late 1940s generally took shape in concert with larger federal efforts overseen by President Miguel Alemán. Visiting the Pacific coastal city of Acapulco in the spring of 1947, Alemán had expressed interest in forming a new tourism promotional agency dedicated specifically to attracting foreign travelers to Mexico.[34]

In January 1948 Alemán met with members of the Veracruz Local Tourism Commission. In their discussion the president indicated his support for local tourism while acknowledging key attractions in the port, such as the island fortress of San Juan de Ulúa, the city's harborfront esplanade (Malecón), and the southern waterfront tourist complex, including the Villa del Mar facility as well as the hotel and nightclub at Mocambo Beach farther south. Still, as Alemán and the others observed, additional investment was needed.[35]

With limited funds, local tourism industry heads focused on facilitating domestic tourism. Hotel, restaurant, and transport opera-

26. Advertisement in *El Dictamen* for Mexico City–Veracruz bus schedules and ticket prices, ca. 1950. Andrew Wood collection.

tors were encouraged to keep prices affordable. State tourism agent Francisco García Marchena again took the initiative in promoting tourism in Veracruz by organizing inexpensive excursions to the port from a handful of locations, including Mexico City, where travelers would depart on Friday night, arrive in the port on Saturday, and stay at the Hotel Villa del Mar. Subsequently, tourists could enjoy visits to neighboring towns of Boca del Río and Alvarado while also circulating freely in the port.[36]

The Carnival festivities in 1948 saw an estimated fifteen thousand tourists come to the port of Veracruz from areas around the state and elsewhere in central Mexico. At the height of the nine-day event, *El Dictamen* proclaimed that "hotels and guesthouses are filled to capacity while many in town are staying in private homes."[37] Once the event was over, city officials estimated that tourists had spent "more than a million pesos," perhaps even double this figure. Observers figured that those providing accommodations, as well as owners of restaurants, bars, and cafés, benefited most. Local and regional transport companies also witnessed significantly increased sales, as did Veracruz retailers and other service providers.[38] Again, Veracruz tourism official García Marchena provided comment: "Tourism brings real income to the city," to which he added, "What we need to do is to try and use this revenue for constructive purposes."[39] Of course not everyone agreed that thousands of tourists flooding into the city was necessarily a good thing, as the occasional satirical cartoon or humorous column published during the height of tourist season would make evident.[40]

Popular opinion notwithstanding, local boosters nevertheless pressed on. An editorial in *El Dictamen* soon noted how the Local Tourism Commission aimed more specifically to capitalize on the rising tide of visitors to the port: "If Veracruz wants to become known as a desirable destination for both national and international travelers . . . investments need to be made to improve accommodations and develop new attractions—not so much for wealthy tourists—but for the type of visitor more often drawn to Veracruz—the average excursionist on a budget."[41] Marketing aimed at attracting domestic tourists of more modest means received further encouragement that summer when local government officials—in accordance with the national tourist administration—announced that hotel, guest house, restaurant, and café owners faced a new set of severe fines if rates and prices deviated significantly from what they typically charged consumers. Detailing what visitors to the port could expect to pay, the policy indicated that one should pay no more than two pesos for a modest-sized *café con leche*, a glass of lemonade, or a pack of cigarettes. "From today," the official state government notice read, "Mexican tourists wishing to spend their vacation in Veracruz can be sure that prices will remain fair and consistent with what one would expect to pay."[42] Apparently, past abuses and a "lack of cooperation" on the part of hotel and restaurant owners at the local level had prompted leaders to act.[43]

In mid-October 1949 a host of government and tourism industry representatives met in Mexico City for a national convention. Addressing the gathering as the keynote speaker, President Miguel Alemán encouraged his audience to think of tourism not only as a business but as a way to unite the Mexican people. "Through tourism," he declared, "our citizens have the opportunity to effectively make use of their civic education and enhance their love of country." Moreover, Alemán suggested that "international tourism should also serve to promote goodwill between peoples of different nations."[44]

Alemán soon visited the port of Veracruz in late November. His agenda included a visit to San Juan de Ulúa to oversee the expansion of maritime trade facilities on the island, inaugurate a new city sewage treatment plant (presumably to redirect wastewater, which once spilled directly into the ocean just south of the city), and meet with the Pro-Veracruz Committee.[45]

The opening of the seaside resort Hotel Villa del Mar on March 30, 1950, provided Veracruz tourism promoters and residents alike with much cause for enthusiasm. With exquisite accommodations as well as a "semiolympic" pool and nightclub on the grounds, the new hospitality complex was poised to do a brisk business. For the inaugural evening, mambo singer Yeye with the Larry Sonn Orchestra performed to a sold-out audience. The next night renowned

27. Ignacio Zaragoza, *Hotel Villa del Mar Swimming Pool*, 1950. Andrew Wood collection.

Veracruz vocalist and recording star Toña la Negra was also warmly received.[46] Subsequent shows would include other entertainment luminaries, such as Sofía Alvarez and the group Los Angeles del Infierno. Overall, the addition of a first-class hotel and recreational facility to the Veracruz waterfront seemed to provide industry observers reason to be hopeful about the future of local tourism.

Urban infrastructural enhancement received a boost in November 1951 when local business leaders donated funds for improved and more attractive lighting along the Malecón del Paseo. At the same time, however, officials continued to struggle with an "engineering problem" (*un cierto error técnico*) that routed city sewage runoff uncomfortably close to the Club de Regatas recreation facility. Talks with members of the national and local tourism agencies and public health agents, as well as the Federal Improvement Association, to resolve the matter were said to be ongoing.[47]

Competition with Acapulco for tourist revenue increasingly became a topic of discussion in Veracruz. Journalist Miguel Rodríguez Jiménez complained about the fact that the Pacific coastal resort received significantly more attention from the federal government, as well as coverage in the national press, than did the

28. Female champion, Veracruz Deep Sea Rodeo, 1948.
Photographer unknown. Andrew Wood collection.

port of Veracruz. According to certain sources, Rodríguez Jiménez observed, "Acapulco plays host to the finest hotels, beaches, and fresh seafood." Nevertheless, "let us not forget," he suggested, "that one can find fantastic places for swimming and dining just south of the port in Boca del Río."[48]

In addition to community efforts intended to refashion Veracruz as a tourist draw, local attention increasingly turned to the colonial fortress San Juan de Ulúa, just across the Veracruz harbor. No longer functioning as a major site of national defense, the aging complex could, in the opinion of many, potentially serve as a historic attraction and social facility. Along these lines navy captain Francisco Mancisidor would soon encourage city officials to convert San Juan de Ulúa into an official, full-time heritage site that could welcome out-of-town visitors by hosting gatherings, concerts, festivals, and tours while also featuring information on prominent individuals and events from the Veracruz past. In the coming years his suggestions would gradually be realized.[49]

Semana Santa

Given the hopeful outlook but also the formidable challenges modern tourism faced in Veracruz, the most promising example of potential industry success came during the busy spring vacation season around Semana Santa, a time when an estimated twenty-five thousand people visited the port and its environs.[50] Coverage of the national vacation week in *El Dictamen* was typically exuberant, with many photos of large crowds (estimated at nearly fifteen thousand) enjoying the beach facilities at Mocambo and Villa del Mar.

Proudly describing the joyful, crowded scene, *El Dictamen* declared that the city had become "the capital of national tourism," where "people of all classes gather" and make the seaside resort a happy and truly a "democratic center of outdoor recreation." Accommodating the many visitors from across the nation (families from Guadalajara, Monterrey, Puebla, and beyond), the local tourism commission began organizing swimsuit contests, dances, concerts, tertulias, and even a bike race to be held at various sites throughout the city. For the 1949 Semana Santa festivities, it was figured that visitors helped generate an estimated 1.5 million pesos in revenue.[51]

An editorial published in *El Dictamen* following the Semana Santa in 1951, "The Proof of Tourism," further extolled the virtues of Veracruz as a site of domestic leisure travel. Despite a cold front that had moved in at the close of the spring vacation, the beaches stretching south along the newly developed Manuel Ávila Camacho Boulevard provided visitors a highly "captivating tropical panorama" of "pleasant weather, gardens and seaside views." The writer confidently asserted that recent public works investments and refashionings (*obras de ornato*) had made the city "beautiful and modern" for both residents and visitors alike. Accommodations and service in Veracruz had indeed come a long way from the old days when residents scrambled to play host to visitors by "improvising places to stay and serving meals in their own homes." "Today," it was observed, "tourist demand has been met in a manner suitable for those coming to enjoy the city and its many attractions."[52] Veracruz, in other words, attracted not so much the high-class international traveler but domestic tourists of more modest means.

Another editorial assumed that tourism was good for all and therefore encouraged residents to cooperate with the ongoing effort to expand the local industry. "What can one do to help improve the trade?" the writer asked. "Fix up one's house, train your employees to anticipate what visitors may want," and "if veracruzanos can meet the needs of tourists," the author of this optimistic piece figured, "the hospitality trade will surely grow."[53]

Information collected by state tourism officials for July 1951 suggested that efforts to promote tourism in Veracruz via print, as well as radio and film, were paying off. An article on tourism marketing being undertaken indicated that some 10,318 tourists had visited the city of Veracruz that month (July) alone.[54] Other reports revealed that tourist traffic had more than doubled from the previous year (nearly 130,000 visitors to the port from January to September 1951). The success or failure of the local tourism trade, according to one state representative, Lieutenant Colonel Domingo G. de los Llanos, depended not exclusively on government or private industry but perhaps just as importantly on the manner in which hosts responded to guests visiting Veracruz. "Whether treated well or poorly during their stay by hospitality employees and the public at large," he wrote, "all tourists share their experience with those they come in contact with back home." Everyone therefore had a "moral obligation," los Llanos asserted, to help ensure that visitors have "a most pleasant stay while in Veracruz."[55] While not an unreasonable proposition, perhaps the lieutenant colonel's remarks indicated that economic, strategic, and physical limits to the city's potential transformation into a tourist destination had been reached. Once again, the occasional "Pepe Peña" satirical column published in *El Dictamen* suggested that at least some local residents had grown quite weary of the annual "invasion" of pleasure-seeking out-of-towners every Easter.[56]

In the interest of consolidating industry efforts, Veracruz state governor Marco Antonio Muñoz proposed that all state agencies working to promote, manage, and administer tourism in Veracruz be brought under his charge beginning in January 1952.[57] Ensuing preparations for Semana Santa 1952 subsequently included an added "folkloric" presentation as part of the program, with a number of *conjuntos de huapango* from Tlacotalpan, Alvarado, and the port of Veracruz. Further enhancements to the weeklong calendar included various *fiestas aquáticos* that no doubt would be of "enormous interest for visitors" to the city.[58] With nearly sixty thousand domestic tourists estimated to be traveling throughout the country during the Semana Santa holiday period, the hospitality industry in Veracruz again rightly figured that, next to Acapulco, their city would rank as the most popular in the nation.[59]

Conclusion

In the immediate postwar years positive indications of Mexico's expanding domestic tourist industry could easily be found during Holy Week in the port of Veracruz. There, on the sandy shores of the Gulf of Mexico, thousands gathered for fun and relaxation. Spending time away from home with family and friends, these Mexican tourists contributed significantly to domestic tourism industry development at a time when others focused more on drawing foreign tourists.

In considering the history of travel in Latin America, it is important to keep in mind that domestic tourists constituted an important constituency for a wide range of industry stakeholders as they set about the task of establishing the basic infrastructure, services, and hospitality culture found today across the hemisphere. Not only in Mexico but also in other Latin American and Caribbean nations, as well as in the United States and Canada, the late 1940s and 1950s represented a critical time for industry growth.

Notes

Epigraph: "Veracruz será un centro de turismo internacional: Tiene condiciones ideales para cualquier visitante," *El Dictamen* (Veracruz), September 7, 1946.

1. In addition to Christmas, New Year's, and Three Kings Day (Epiphany), Carnival, Semana Santa, and summer vacation during the first weeks of August, a popular time for travel to Veracruz, Mexico, also included Day of the Dead in early November. See, for example, "Cuando menos cinco mil turistas vinieron de vacaciones al puerto," *El Dictamen*, November 3, 1948, as well as "Millares de visitantes de vacaciones en el puerto: Los días de Muertos trajeron gran afluencia de turistas," *El Dictamen*, November 3, 1951. Overall, an estimated fifteen thousand to twenty thousand travelers were said to have made their way to the port in 1951—a new high. "De quince a viente mil turistas nos visitaron: Nunca antes en estos dias se habia visto tal afluencia," *El Dictamen*, November 4, 1951; "Vinieron por tren cinco mil turistas," *El Dictamen*, November 6, 1951.

2. "Veracruz debe ser habilitado como puerto aéro internacional: Consideraciones sobre el future de las comunicaciones con el extranjero," *El Dictamen*, July 11, 1949. See also "La habilitación de Veracruz como puerto aéreo," *El Dictamen*, August 10, 1949.

3. The considerable success of these events was celebrated in the pages of the local newspaper *El Dictamen*. Several internationally known sport fishing aficionados participated in these tournaments, including—for the 1948 tournament—Dick Tronson, Leroy H. Dorsey, R. W. Winter, James Swaam, and Virgil Lehman. "El viernes se inicia en este puerto el III Rodeo Internacional de Pesca," *El Dictamen*, April 7, 1948. For a later event, see photos of tournament winners Ernesto Sanders and "Señorita" Chuchy Carvallo de Villegas (looking much like Marilyn Monroe) taken at the Boca del Río Fishing Club, in *El Dictamen*, April 29, 1952.

4. Fabulous photos of Veracruz history and culture are available in García Díaz, *Puerto de Veracruz*.

5. "Preparan el programa para las fiestas de Semana Santa: Calcúlase que 15 mil turistas visitarán el puerto en esos días," *El Dictamen*, March 28, 1952; "25 mil turistas se calcula que visitaron este puerto: Ratifica Veracruz su lugar como centro de atracción," *El Dictamen*, April 13, 1952.

6. *La Revista Nacional de Turismo* 1, no. 1 (June 1930).

7. Continued organization and promotion of annual late winter Carnival festivities in the city beginning in 1925 drew a growing number of visitors from central Mexico and beyond. On the history of Carnival, see Wood, "Introducing La Reina del Carnaval."

8. Rose, "Veracruz," 48.

9. Rose, "Veracruz," 48.

10. For local Veracruz reporting, see "Veracruz ocupa un lugar de importancia en el mundo: 46 barcos de altura traen cada mes 80 mil toneladas de carga," *El Dictamen*, April 18, 1952; and "Más de veintodos milones para obras portuenses," *El Dictamen*, March 3, 1952.

11. Rubén C. Navarro, "Veracruz será un centro de turismo internacional: Tiene condiciones ideales para cualquier visitante," *El Dictamen*, September 7, 1946.

12. "Bienvenidos a Veracruz, amigos hoteleros!," *El Dictamen*, October 24, 1946.

13. "La cooperación entre hoteleros será beneficiosa," *El Dictamen*, October 24, 1946.

14. "Quinta Convención Nacional de la Asociación Mexicana de Hoteles," *El Dictamen*, October 24, 1946.

15. "Transcendencia de la V Convención de hoteleros: Veracruz, simbolo del futuro de nuestro pais," *El Dictamen*, October 24, 1946.

16. "Magnifico resultado alcanzó la V Convención de hoteleros," *El Dictamen*, October 27, 1946.

17. "8 milliones ha dejado el turismo en este año: Han venido a Veracruz no menos de 50 mil visitantes," *El Dictamen*, September 27, 1946.

18. "Veracruz en el turismo mundial," *El Dictamen*, January 31, 1947.

19. Felipe Lagardy, "La feria de la Candelaria en Tlacotalpan," *El Dictamen*, February 4, 1947.

20. "Miles de turistas vendrán a las fiestas de Carnaval: Hoy en la noche se elegirá rey feo en la Plaza de Armas," *El Dictamen*, February 8, 1947.

21. "Miles de turistas vendrán a las fiestas de Carnaval."

22. "Gran impulse para fomentar el turismo en esta ciudad," *El Dictamen*, December 6, 1946.

23. "Antes de otra cosa, se higienizará a la ciudad: Prioridad al servicio del agua potable y el drenaje," *El Dictamen*, May 16, 1947.

24. "Se necesitan 5 milones para mejorar la ciudad: Veracruz debe estar a la altura de su categoria," *El Dictamen*, n.d., 1947.

25. "Causas por las que ha disminuído el turismo," *El Dictamen*, July 29, 1947.

26. "Causas por las que ha disminuído el turismo."

27. Quoted in "El turismo puede dejar 40 millones a Veracruz: Sin embargo, indebidamente esa industria está viniendo a menos," *El Dictamen*, July 30, 1947.
28. Quoted in "Urge incrementar la corriente turística," *El Dictamen*, August 5, 1947.
29. "Porqué huye el turismo de Veracruz," *El Dictamen*, August 10, 1947.
30. Alfonso Valenzuela, "Son tres mil tuberculosos," *El Dictamen*, August 18, 1947.
31. "Efecturán una campaña de desinfección y desratización," *El Dictamen*, April 10, 1951. Mention is made in this article of obligatory sanitizing of hotels, schools, public dormitories, rent houses, vehicles, barbershops, and hair salons, as overseen by the Secretaría de Salubridad y Asistencia. Further provision was to be made in many instances for the circulation of air so as to dry out overly humid, moldy areas. See also "El miércoles inicia la gran campaña de limpieza: Propósito de las autoridades de que la ciudad sea positivamente limpia," *El Dictamen*, June 24, 1951; "Hoy se inicia la campaña la limpieza en las casas: Salubridad espera cooperación en la lucha contra la poliomyelitis," *El Dictamen*, June 27, 1951; and "Un Veracruz más limpio" (editorial), *El Dictamen*, July 2, 1950.
32. In mid-July 1946 residents had begun organizing over the issue of housing. An association, the Liga de Control y Resistencia Inquilinaria, had begun a grassroots effort to increase housing stock and control rents. Property owners, along with the mayor and the governor of the state of Veracruz, were called upon to help improve the situation. See "Editorial: Veracruz sin habitaciónes," *El Dictamen*, December 17, 1946; "Se constituirá la Liga de Resistencia Inquilinaria: Para tartar los problemas de escasez de viviendas," *El Dictamen*, July 13 1946; and "Se teme que resucite la agitación inquilinaria: Con la creación de la Liga de Control y Resistencia," *El Dictamen*, July 19, 1946. Curiously, organizers, along with the editors of *El Dictamen*, assured *porteños* that protest actions would not be disorderly in the way housing protests had been in the 1920s. "Garantizan el orden en la resistencia inquilinaria: Las manifestaciones de 'no pago renta' no se efectuarán," *El Dictamen*, July 22, 1946. The following month the newspaper reported that efforts to locate inexpensive materials for more affordable housing construction sought out by Adolfo Seeman and associates in Houston, Texas, had not proven worthwhile. "Dificultad para construer las casas baratas," *El Dictamen*, August 1, 1946. In early March 1947 residents then invaded the property of Juan Roldán Romero (at the end of Díaz Mirón Avenue) and quickly built seventy-five small wooden houses. "Illegalmente iban a invader unos terrenos particulares," *El Dictamen*, March 1, 1947. Those taking action claimed that they needed the space to build homes and that the property was otherwise unused. Subsequent coverage of the housing problem came on March 12, when *El Dictamen* reported that some small-property owners (who often rented to tenants) were now paying more for utilities than what they collected in rent. "Desesperada situación de los propietarios de casas: En algunas las contribuciones son más altas que las rentas," *El Dictamen*, March 12 1947.

For discussion of housing protest in Veracruz during the 1920s, see Wood, *Revolution in the Street*.

33. "12 mil toneladas de basure se sacaron de este puerto: Resultado de la campaña de limpieza efectuada en dos meses," *El Dictamen*, September 1, 1951.

34. "El impulse al turismo," *El Dictamen*, May 28, 1947.

35. "La commission de turismo con el sr. president," *El Dictamen*, January 27, 1948.

36. "Hacen vastos planes para aumentar la corriente turística a Veracruz: Se preparán viajes de fin de semana por carretera en carros especiales," *El Dictamen*, February 21, 1948.

37. "Invadido por miles de turistas, Veracruz desbordó toda su alegría en el primer desfile de carnaval," *El Dictamen*, February 9, 1948.

38. "Más de un millón de pesos dejaron los turistas en el ultimo carnaval," *El Dictamen*, February 13, 1948.

39. Quoted in "Más de un millón de pesos dejaron los turistas en el ultimo carnaval."

40. Illustrated columns in *El Dictamen* deliberately poked fun at tourists. See, for example, "La semana festiva: A las playas! por Pepe Peña," *El Dictamen*, March 21, 1948. See also "La semana festiva: Primavera" and "Ensalada turistica" (containing overheard comments and conversation by visiting tourists), also by Pepe Peña, in *El Dictamen*, March 28, 1948.

41. "Industria turística," *El Dictamen*, January 14, 1948.

42. "Medidas para proteger el turismo," *El Dictamen*, July 30, 1948.

43. "Poderoso impulso economico al Comité de Turismo de esta ciudad," *El Dictamen*, November 3, 1948. These same warnings would be reissued over the coming years in anticipation of popular festival and vacation dates. See, for example, "Se evitarán abusos con el turismo," *El Dictamen*, September 1, 1951.

44. Quoted in "Orientacion del turismo," *El Dictamen*, October 21, 1949.

45. "El Presidente inauguró la planta de tratamiento de aguas negras," *El Dictamen*, November 26, 1949; "Visitará hoy el Presidente de la republic las obras de Ulúa," *El Dictamen*, November 26, 1949.

46. Advertisement featuring Toña la Negra and others, as well as the Larry Sonn Orchestra, *El Dictamen*, March 31, 1950.

47. "Poderoso impulso económico al Comité de Turismo de esta ciudad," *El Dictamen*, November 3, 1948. Subsequent management of the beaches would include efforts to slow erosion. See "Quieren impeder la destrucción de las playas," *El Dictamen*, May 13, 1951.

48. Miguel Rodríguez Jiménez, "Acapulco y Veracruz," *El Dictamen*, March 26, 1948.

49. Francisco Mancisidor, "Turismo vs. Marina," *El Dictamen*, July 31, 1948.

50. "Más de veinticinco mil personas visitaron Veracruz en Semana Sta.: Dieciocho mil regresaron por tren y siete mil en autobuses y coches," *El Dictamen*, March 30, 1948.

51. "Para aprovechar el turismo," *El Dictamen*, April 28, 1949. Articles on Semana Santa that ran in the 1950s further demonstrated the mass appeal of the Veracruz beach scene for tourists and the significant revenues tourism produced for the city, as well as the growing program of offerings being developed to entertain visitors. See, for example, "La ciudad se halla llena de visitants y de alegría," *El Dictamen*, March 22, 1951; "Miles de turistas invaden las playas de este puerto," *El Dictamen*, March 19, 1951; "Continúa la avalanche de visitants a este puerto," *El Dictamen*, March 20, 1951; and "Informe de la Oficina de Turismo," *El Dictamen*, March 26, 1951, which reported 10,472 tourists visiting during Semana Santa, with 4,000 in guest houses and 10,000 staying in family homes. See also "Tres millones de pesos dejó el turismo en Semana Santa," *El Dictamen*, March 27, 1951. As before, the local tourism commission was compiling statistics and would be sharing information with national-level industry and government offices.

52. "La pruba del turismo," *El Dictamen*, March 30, 1951.

53. "Cooperación para el turismo," *El Dictamen*, March 21, 1951.

54. "Eficaz propaganda turística se realiza en este puerto," *El Dictamen*, August 14, 1951.

55. Figures and quotations from "Aumento progresivo del turismo en esta población: Desde enero hasta septiembre habían venido 127,837 visitantes," *El Dictamen*, November 10, 1951.

56. See, for example, "Las consabidas vacaciones por Pepe Peña," *El Dictamen*, April 12, 1952.

57. "Serán reorganizadas las dependencias de turismo," *El Dictamen*, November 25, 1952.

58. "Preparan el programa par alas fiestas de semana santa," *El Dictamen*, March 28, 1952; "Es atractivo el programa de festejos para Semana Santa: Bailables folklóricos serán ejecutadas en Plaza de Armas," *El Dictamen*, April 4, 1952.

59. "25 mil turistas se calcula que visitaron este puerto: Ratifica Veracruz su lugar como centro de atracción," *El Dictamen*, April 13, 1952; "Están llenos los hoteles y casas de huéspedes," *El Dictamen*, April 11, 1952.

Bibliography

Berger, Dina. *The Development of Mexico's Tourism Industry: Pyramids by Day, Martinis by Night*. New York: Palgrave Macmillan, 2006.

———. "A Drink between Friends." In *Adventures into Mexico: American Tourism beyond the Border*, edited by Nicholas Dagen Bloom, 13–34. Lanham MD: Rowman and Littlefield, 2006.

Berger, Dina, and Andrew Grant Wood, eds. *Holiday in Mexico: Critical Reflections on Tourism and Tourist Encounters*. Durham NC: Duke University Press, 2010.

Boardman, Andrea. *Destination Mexico: "A Foreign Land a Step Away"; U.S. Tourism to Mexico, 1880s–1950s*. Dallas: DeGolyer Library, Southern Methodist University, 2001.

Clancy, Michael M. *Exporting Paradise: Tourism and Development in Mexico.* Oxford: Pergamon Press, 2001.

Covert, Lisa Pinley. *San Miguel de Allende: Mexicans, Foreigners, and the Making of a World Heritage Site.* Lincoln: University of Nebraska Press, 2017.

Delpar, Helen. *The Enormous Vogue of Things Mexican: Cultural Relations between the United States and Mexico, 1920–1935.* Tuscaloosa: University of Alabama Press, 1992.

García Aguirre, Feliciano Joaquín. "Economía veracruzana del siglo XX." In *Historia general de Veracruz*, edited by Martín Aguilar Sánchez and Juan Ortiz Escamilla, 487–545. Xalapa, Mexico: Gobierno del Estado de Veracruz, Secretaría de Educación, 2011.

García Díaz, Bernardo. *Puerto de Veracruz: Imágenes de su historia.* Xalapa, Mexico: Gobierno de Veracruz/Archivo General del Estado, 1992.

Koutsoyannis, Sofia. "Immoral but Profitable: The Social and Cultural History of Cabarets in Mexico City (1920–1965)." Doctoral diss., York University, 2010.

Merrill, Dennis. *Negotiating Paradise: U.S. Tourism and Empire in Twentieth-Century Latin America.* Chapel Hill: University of North Carolina Press, 2009.

Moreno-Brid, Juan Carlos, and Jaime Ros. *Development and Growth in the Mexican Economy: A Historical Perspective.* New York: Oxford University Press, 2009.

Rose, Halleck L. "Veracruz: A Possible Sea-Side Resort." *Real Mexico* 2, no. 8 (1933): 48–51.

Ruiz, Jason. *Americans in the Treasure House: Travel to Porfirian Mexico and the Cultural Politics of Empire.* Austin: University of Texas Press, 2014.

Saragoza, Alex. "The Selling of Mexico: Tourism and the State, 1929–1952." In *Fragments of a Golden Age: The Politics of Culture in Mexico since 1940*, edited by Gilbert Joseph, Anne Rubenstein, and Eric Zolov, 91–115. Durham NC: Duke University Press, 2001.

Vanderwood, Paul J. *Satan's Playground: Mobsters and Movie Stars at America's Greatest Gaming Resort.* Durham NC: Duke University Press, 2010.

Wood, Andrew Grant. *Agustín Lara: A Cultural Biography.* Oxford: Oxford University Press, 2014.

———. "Introducing La Reina del Carnaval: Public Celebration and Public Discourse in Veracruz, Mexico." *The Americas* 60, no. 1 (2003): 87–107.

———. *Revolution in the Street: Women, Workers and Urban Protest in Veracruz, 1870–1927.* Lanham MD: Rowman and Littlefield, 2001.

8

The Hotel Casino Project That Put Ecuador's Tourism Hopes on Pause

KENNETH R. KINCAID

On January 8, 1959, hundreds of Ecuadorian indigenes from the *comuna* Pucará Bajo de Velásquez, in the canton of Otavalo, mobilized to stop municipal plans to construct a hotel-casino on the shores of their revered Imbakucha (Lake San Pablo). While members of the planning commission met at the lake's pier to discuss the logistics of the project and attempt to get signatures conceding some of the comuna's lands, reportedly more than five hundred indigenes descended upon them with sticks raised, stones aimed, and fists clenched. The delegation intended to explain that the project was imperative since Ecuador had been selected to host the forthcoming Eleventh Inter-American Conference and that this project was needed to promote tourism to Imbabura Province. The insurgents, though, had grown weary of attempts by the city council to usurp the lands along the lake, efforts that spanned more than four years, and they had come to the conclusion that the city council had decided to begin the project with or without the sale of lands. Poised to attack, the Pucareños denounced the municipal government and vehemently insisted that the city had no right to take their land—or their water either. Local mestizos, believing that the commission was in peril, armed themselves and began firing indiscriminately into the indigenous masses. The arrival of the armed guards did not quell the violence, and the fighting lasted for another hour. Finally, after several deaths, injuries, and arrests, the residents of Pucará capitulated. In total, five indigenes were killed, hundreds more were injured, and scores detained. The city council defended its actions and those of its mestizo population, claiming that they had acted only in self-defense. In an attempt to put the ordeal behind them once and for all, Victor A. Jaramillo, city council president, concluded that the events of that day were nothing more than the product of a "misunderstanding between indigenes and the city."[1]

Despite the tragedy, city leaders insisted that construction of the hotel carry on as planned as national leaders continued to prepare for the Eleventh Inter-American Conference. Critics, however, argued that it would be untimely to hold the event after the massacre and took steps to stop the project. The attempt by Pucará Bajo's indigenous population to defend its land and access to water struck a nerve in the national psyche, and for the next several weeks Ecuador's media published scathing attacks on Otavalo's "racist" city council and questioned the motives behind the decisions and actions that had led to the bloodshed.[2] The events that unfolded in those fateful days around Lake San Pablo and the city council's chauvinistic attitude toward the indigenes of the lake basin were reminders to national policy makers of the need to create legislation that would help eradicate some of the inequities in Ecuadorian society and address the need for fundamental change in land and water laws.

The incident at Lake San Pablo in 1959 provides a reminder of the challenges of developing a tourist infrastructure in countries where the benefits of progress are unequally shared. This statement is particularly true in areas with large indigenous populations and sacred landscapes.

The 1959 Inter-American Conference in Quito, Ecuador

By acclamation at the 1954 Inter-American Conference in Caracas, Venezuela, delegates selected Quito, Ecuador, as the host city for the Eleventh Inter-American Conference. Since the charter of the Organization of American States (OAS) instructed its principal organ, the Inter-American Conference, to convene periodically (every five to six years) and since it had done so in 1948 in Bogotá and again in 1954 in Caracas, it planned its next meeting for 1959 in Ecuador's highland capital.[3]

For this small Andean nation of about four million inhabitants, nestled between Peru, Colombia, and the Pacific Ocean, the conference was an opportunity to take center stage among members of the OAS. Not only would the Inter-American Conference forum allow Ecuador the opportunity to address recent disputes with some of its neighbors, it would also give this country a chance to develop its tourist industry and showcase the wonders of the coast, the highlands, and the Amazon.

Upon accepting the nomination to host the Eleventh Inter-American Conference, President José María Velasco Ibarra, who at the time was serving a third nonconsecutive term as president of Ecuador, used his power of executive decree to appoint a *junta coordinadora permanente*, a permanent coordinating board charged with assisting him in preparations for the event.[4] Included in the nine-member junta were the vice president of the republic, the mayor of the capital city of Quito, and three ministers, one of whom was the minister of *previsión social* (social welfare). In 1955 and 1956 Velasco addressed some of the financial considerations of the conference by issuing executive decrees that capped spending at 150 million sucres on construction and operating expenses related to the conference. He listed a legislative palace, a presidential house, and infrastructural projects in Quito as essential to the success of the conference. To pay for the project, Velasco committed 10 million sucres from the annual national budget.[5] He also imposed a 0.5 percent tax on cash income generated by almost all of the nation's public and private institutions.[6] Despite these measures, construction and organization advanced haltingly, and neither the Junta Coordinadora Permanente nor the Velasco administration provided sufficient guidance or planning prior to the close of his term in 1956.[7]

Succeeding Velasco as president was his minister of the interior, Camilo Ponce Enríquez. Ponce inherited an economy that stood

29. Map of Ecuador. Wikimedia Commons.

on the precipice of bankruptcy. The new chief of state immediately replaced Velasco's 1956 budget with his own, slashing federal spending and implementing a meaner, leaner government. His advisors suggested that Ecuador abandon plans to host the Inter-American Conference, as it was clear that the country did not have the resources to cover costs. For Ponce and his supporters, however, canceling the conference was something that no one truly wanted to entertain seriously. Such an act, it was argued, would discredit the Ponce administration, "as Ecuador's hemispheric prestige would have been dealt a major blow."[8] Rather, Ponce kept hopes for the conference alive but reduced the budget by almost 10 million sucres while borrowing $2.5 million from the International Cooperation Administration of the U.S. government and $8 million from London Bank.[9] Ecuador's ambassador to the United States, José R. Chiriboga, played a significant role in negotiations with Assistant Secretary of State Thomas Mann to secure the loan, pointing out that the creation of a tourist hotel with two hundred to three hundred rooms would "promote the development of tourism which would in turn benefit the entire economy of Ecuador."[10]

The austerity measures imposed by Ponce seemed to pay off, as Ecuador's national budget seemed to normalize. However, construction of hotels and other amenities needed for an international conference lagged. Budgetary problems and the inability to import the necessary materials forestalled the completion of Hotel Quito, the Palace of Government, the Legislative Palace, university residences, and the Quito and Guayaquil airports until after the initial opening date of the conference. Even minor repairs of capital streets and other components of the urban infrastructure did not get *cabildo* (town council) approval until as late as December 1959.[11]

Although cancellation of the conference was out of the question for Ponce's government, postponement of it was not, and in December 1960 Ecuador's foreign minister announced that as a result of "inevitable delays . . . perhaps a postponement of the Eleventh Conference would be prudent."[12] One of the reasons that Ecuador remained steadfast in its intent to host the Inter-American Conference despite myriad obstacles was its desire to use the event to garner international support in repudiating the Protocol of Peace, Friendship, and Boundaries Treaty signed by the Conference of Foreign Ministers in Rio de Janeiro in January 1942 (hereafter, Rio de Janeiro Protocol) that established Peru's and Ecuador's Amazonian territorial boundaries. Ecuadorians harbored deep-seated ill will toward all those who had signed the treaty or made the treaty possible, from their very own President Carlos Arroyo del Río to the Peruvian-Ecuadorian Boundary Commission and the treaty's "friendly" guarantors—the United States, Argentina, Chile, and Brazil. Multiple times during the territorial demarcation process Peru and Ecuador asked the guarantors to settle disputes. Tensions between Peru and Ecuador escalated even more in 1954 and 1955 as a result of actions taken by both countries, including charges of Peru mobilizing troops along the border, violating Ecuadorian airspace, and granting mining concessions to a Canadian company in the disputed area and of Ecuador preparing for military action against Peru. Both countries used the organs of the Organization of American States, including the Inter-American Conference and the Council of Foreign Ministers meetings, to attack the other.

When Ecuador was awarded the Eleventh Inter-American Conference, Peru suggested that it might boycott the meeting if it did not receive assurances from the guarantor nations and Ecuador that the Rio de Janeiro Protocol would not be discussed. In order to appease Peru and the guarantor nations, Ecuador's President Ponce called for a truce that would last for the duration of the conference. Ponce's foreign minister, Carlos Tobar Zaldumbide, supported the president's position, arguing that Ecuador's decision-makers needed to avoid nationalist demagoguery and think internationally. Ecuador's Congress, which was predominantly *antiponcista* (anti-Ponce), criticized the accord and called for the resignation of Zaldumbide. The matter became more contentious when former Ecuadorian presidents José María Velasco Ibarra and Galo Plaza both took stands in opposition to that of the Ponce administration and suggested that the conference would be an ideal forum in which to discuss the treaty. Not only did it appear that former presidents were attempting to sabotage the meetings, but the Ponce

30. Otavalo Plaza. Wikimedia Commons.

government was also having difficulty putting together a committee to represent Ecuador at the conference. Indeed, charges that Ecuador's own internal political chaos was behind the problems faced by the conference resonated well beyond Ecuador's borders. Even Ecuador's Communist Party was named as a possible culprit for hijacking the meeting.

Otavalo and the Eleventh Inter-American Conference

The mistakes and miscalculations that forced Quito to postpone the international conference also plagued Otavalo's preparations for the event. However, unlike in Quito, where the consequences were limited to economic and political fortunes, in Otavalo the mistakes and "misunderstandings" surrounding the preparations for the Eleventh Inter-American Conference assumed a tragic dimension as several people were killed, hundreds more injured, and scores more imprisoned. For many, the failure to deliver the conference as planned and the massacre of several indigenes at a potential hotel site reflected not only Ecuador's incapacity to join the ranks of the fraternity of nations in the OAS that had fulfilled their conference-hosting responsibilities but also its archaic social structure, one that continued to deprive native peoples of the right to their own land and labor.

From the perspective of pure aesthetics, the decision to host part of the Eleventh Inter-American conference at Lake San Pablo in Otavalo, Ecuador, could not have been better. The lake sits at the base of Imbabura Volcano and receives its waters from occasional snowmelt and streams that descend from this mountain as well as other mountains, such as Cusin and Mojanda. The lake is deep and provides ample space for water sports. Moreover, national and municipal planners had previously introduced exotic species, such as eucalyptus trees, to "beautify" the environs of the lake, and trout, to promote sport fishing for tourism and trade with Colombia.

In addition to stunning lakes and awe-inspiring mountains, the canton of Otavalo is home to dozens of indigenous communities, many of whom reside along the shores of Lake San Pablo. Of the 36,653 residents of Otavalo counted in the 1962 national census, more than 30,000 were indigenous.[13] Ethnic relations between whites/mestizos and indigenes varied based on locality. Whereas indigenes and whites/mestizos typically had amicable relations in and around the urban areas of Otavalo, around Lake San Pablo those relations were often more contentious as conflicts over land and water rights frequently escalated.

Nevertheless, Otavalo's city council and the conference planners embraced the opportunity to show off their city and lake and hoped that the Eleventh Inter-American Conference would help stimulate the tourism industry. Some even envisioned Otavalo as taking a primary role in the conference as a result of the slow rate of construction and preparation in the capital. Taking the lead in Otavalo's participation in the Eleventh Inter-American Conference were the Asociación 31 de octubre (October 31 Association), an organization of Otavalo-born whites and mestizos residing in Quito, as well as the president of the city council, the principal of the Otavalo's National Preparatory School, and the senator representing the province of Imbabura, Victor Alejandro Jaramillo.

In the days preceding the tragedy, open town hall meetings (*cabildos abiertos*), general commissions, and popular assemblies convened frequently in order to promote the building of a tourist hotel along the shores of Lake San Pablo. Proponents argued that upon completion the hotel would provide a boon to the region's economy. However, the commissions also expressed concern that the realization of the hotel faced an uphill battle since the city did not have the land on which to build it.[14] Others countered that the opportunity for finally constructing the long-awaited hotel had arrived, since Ecuador was hosting the Eleventh Inter-American Conference and the country needed hotels for the delegates. The local planning commission, they argued, needed to act, because failure to take advantage of this opportunity would doom any hope for economic progress for Otavalo and, more generally, the province of Imbabura.

The previous municipal government had advanced the project considerably by creating and approving studies and maps, acquiring funds, and laying out contracts with a U.S. construction company. However, it failed to complete the most important task: acquiring the land needed for the project. They took it for granted that this problem would be easy to resolve. As it turned out, the purchase of the land for the tourism hotel became a major sticking point, and the optimism of the advocates for a hotel-casino died when on two occasions the commission created to negotiate the land transfer was met with aggression by the indigenes, who, in their majority, rejected any deal-making or territorial sales.

Jaramillo insisted that the leaders of Pucará Bajo were being influenced by outside agitators who persuaded them that the commission was going to take their lands without payment. Moreover, according to Jaramillo, the agitators had convinced the indigenes that the city was not only going to take thirteen thousand square meters of land from them but that they actually had planned to take all of their lands adjacent to the lake.[15]

Noted Otavalo specialist Álvaro San Félix reported that "the offer and the demand failed as a result of the impossibility of a real understanding" between the two sides: "The fact that the Municipality had offered to pay the indigenous landowners double the value of the lands and give them an additional 30,000 sucres, or give them, in exchange, double the amount of land" failed to convince the indigenes of Pucará Bajo, as "they did not want to cede even an inch."[16]

According to Jaramillo, in numerous council sessions and by radio he put forth the possibility of selecting another site, since the indigenes did not want to give up their land. For many of the council members, however, "things were easy and they insisted on obtaining that land." As such, they formed a commission to negotiate the transfer of indigenous lands and included church officials in this group, in order to exploit the respect and high spiritual standing of the clergy over the native landholders. Among these church leaders was Father Polibio Andrade, prior of the Franciscans and principal of Solano Preparatory School. However, even with the support of the Catholic Church, the commission failed to elicit any type of favorable response from the indigenes of Pucará Bajo de Velásquez.[17]

Some supporters of the project began to entertain the idea of building the hotel on lands that the Ministry of Social Welfare had already declared to be of public utility or building it on those lands that had already been obtained legally by whites, such as in Araque, where Dr. Isidro Ayora had already offered to sell the land to the municipal government. However, these were not the lands that the municipal government wanted. Despite Jaramillo's insistence that it was particular members of the commission who wanted to acquire the ideal spot for the tourism hotel, in some of his later writings he identified a different area, the shoreline adjacent to the Pucará Bajo de Velásquez community, as the ideal site for a tourism complex. The popular assemblies reconvened, and speeches in favor of obtaining the lands of the indigenes won out over those who preferred a nonconfrontational solution, and the assemblies resolved, albeit cautiously, to attempt again to persuade the Pucareños to sell their lands.

On January 8 members of the municipal commission for acquiring land from the indigenes of Pucará Bajo met. The group consisted of the city attorney (*procurador síndico*), Pedro Alarcón; the mediation lawyer, Dr. Alfredo Rubio; the police lieutenant of San Luis de Otavalo (*teniente politico*), Alfonso Paredes; and the secretary to the police lieutenant, Rafael Barahona. In addition, Doña Matilde Auz, who knew personally the indigenous landowners and an indigenous leader known as the *ñaupador*, also accompanied the commission in the hope of visiting the landowners in their houses before taking them to the pier, where, if all went as planned, the purchase would peacefully take place.

According to Rubio, when the messenger began his ascent up the hill to contact the landowners, he heard whistles and calls from hut to hut. It was apparent to him that Pucará Bajo had prepared for the moment that the commission returned to their town. The signals put on pause the celebratory mood of the community, which was in day two of José Guamán and Barbarita Males's wedding party, and within a few minutes multitudes of indigenes began to descend the hill toward the pier. The commissioners, upon seeing the mass of indigenes, took refuge in a house at the pier. The teniente político, Alfonso Paredes, alarmed at the site of hundreds of indigenes heading for them, ran toward a nearby municipal dump truck and asked the driver to take him back to Otavalo so that he could report the uprising.

Paredes arrived in Otavalo erroneously conveying news of a massacre of whites at the hands of the indigenes of Pucará Bajo de Velásquez. Later he reported that "the Indians were attacking members of the City Council Commission . . . with rocks and sticks." Paredes decided to go to the Guardia Civil (Civil Guard) for help, but he could not find anyone. Next, the teniente político went to the Comisario Nacional (National Police) and was told that not a single officer was available at that moment and that he should ask for help from the Empleados y Guardias de Estancos (Employees and Guards of the State-Run Tobacco and Alcohol Store). According to both Jaramillo and Paredes, all of this took place without the knowledge of Jaramillo, who arrived later at Lake San Pablo.[18]

In Jaramillo's version of the events, he points out that when he saw Paredes, the teniente político's face was anguished and he was almost hoarse when he told him that "the Indians are killing the commissioners in the lake." Jaramillo explained later that since Paredes occupied a position of authority in Otavalo, he viewed the officer's statement as credible. Stunned by the magnitude of what he was hearing, Jaramillo accepted Paredes's account, left his principal's office, and communicated the information to two professors, one of them his brother Estuardo and the other, Lauro Samaniego. Panic-stricken, Jaramillo interrupted classes and explained to the students that the lives of the commissioners and of a professor were in danger.[19]

Jaramillo gave permission to the students to board the municipal dump truck and go with the driver to the site of the uprising. Jaramillo hoped to use the arrival of the students in the truck as a distraction, in order to extricate the besieged members of the commission from the house where they had found refuge. By the time the students arrived, the guards from the tobacco and alcohol store had already arrived.

However, neither they nor Jaramillo saw the carnage that Paredes believed he had seen; there was fighting for sure, as the white townspeople from Eugenio Espejo went to the pier to prevent the indigenes from attacking the commission; however, there were no white commission members horribly beaten, no cadavers strewn about on the ground, no blood drenching the earth. Nevertheless, since the guards believed that the onslaught was imminent, they began to shoot. According to one student, "After the commission members were able to leave the pier, the person who was in charge of the Estanco guards ran toward a small wall and aimed carefully, shot at what later he saw was a group of Indians who had climbed the trees. One fell to the ground, dead. Later they (the members of the Estanco) began to climb up the hill, shooting at the mass of people trying to escape."[20]

Jaramillo arrived at the lake accompanied by employees and members of the municipal band, who had suspended their rehearsal in order to defend the planning delegation. According to a statement signed by several Otavalo citizens, at the time of Jaramillo's arrival hundreds of indigenes were attacking the white residents of Eugenio Espejo who had mobilized to save the commission. Moreover, gunfire had already broken out. Jaramillo got out of the car and, hurrying ahead in the direction of the shooting, shouting with all of his energy and repeating several times "not to shoot." Jaramillo in his statement at the subsequent hearings explained his actions:

> Upon my arrival I saw the use of arms and I heard the sound of shots. The rocks thrown from the other side whistled all around me. A feeling of courageous passion for the combatants led me to fling myself out of the car, while the commission was retreating due to the brutality of the attack, in order to ask insistently that those who were armed, even while risking my life, cease the fratricidal battle and that all of us return prudently to the town of Eugenio Espejo.... My pleas were not heeded. In fact, many rose up against me, twice the combatants reproached me and my pacifism, in circumstances that nobody can really understand[,] the imminent danger that we were in[,] the furious whirlwind that was engulfing us.[21]

While the fighting continued, some of the students who had gotten out of the dump truck and begun to throw rocks at the native residents now found themselves the target of the indigenes' fury. They retreated to the truck and left before the mass of indigenes could overtake them. Two students were gravely injured in the ordeal.

By the time the fighting had subsided, five indigenes had been killed: José Guamán, Francisco Guamán, José Velasques Cahuasquí, José Andrango Velasques, and sixteen-year-old Ignacio Velasques Guambrango. Countless other people were injured, and Otavalo's prison was filled with more than thirty indigenous prisoners, including the bride and groom whose wedding celebration was disrupted by the arrival of the municipal planning commission.

The political fallout was immediate. Critics of Jaramillo, who was a *poncista*, or supporter of President Ponce, had two reasons for their attacks on the senator: first, his decision to use students to put down the uprising and second, the use of armed force against an unarmed people. The press, the radio, the editorials in the newspapers and magazines, and the caricaturists attacked Jaramillo harshly. Colleagues in education called for sanctions against the rector and removal from his post. Not only was Jaramillo responsible for the fact that two students had been injured in the uprising, but these young people had been "put in opposition to a people that deserves ... the utmost respect."²² The Ministry of Education created a commission to investigate Jaramillo's actions. The Federation of Workers of Pichincha (Federación de Trabajadores de Pichincha, FTP) defended the decision of Pucará Bajo to resist the municipality's efforts to expropriate these lands, arguing that "they are a legacy of their ancestors and it is there that their children were born and raised."²³ Moreover, the FTP protested "the armed and violent intervention of the Police against the indigenous comuna of Pucará Bajo, whose members have died as a result of defending their lands against the unjust expropriation proposed by the Municipal Government of Otavalo." The Junta Provincial de Pichincha of the Liberal Radical Party also condemned the "double crime provoked by the Senator of Imbabura."²⁴

On January 20, 1959, Dr. Alfonso Mora Bowen, attorney general of the nation, arrived in Otavalo as special investigator in order to oversee the judicial process and discover the authors, accomplices, and accessories to the crime. He found that the civil guards and the tobacco and alcohol store guards involved in the matter had been relocated quickly to other cities. Mora Bowen stayed in Otavalo several days while trying futilely to get statements from the indigenes, but they did not trust him or the system and were unwilling to present complaints.²⁵

The exception was the grandmother of Ignacio Félix, the sixteen-year-old who was killed. In her statement she contended that whites were shooting at indigenes as they tried to escape to their houses or climb trees. With the Pucareños on the run toward the hills, whites took to looting houses, taking everything they could, including *huacas* and money.²⁶ She pointed out that her grandson was alive the last time she saw him, but he had been kicked repeatedly before being led away.²⁷

On January 13 Carlos Bustamante Pérez, head of the Ministry of Government, declared after a meeting with the commissioners and delegates, "There has taken place a misunderstanding among all. The indigenes do not understand the sensible intentions of the Municipality; the Municipality did not understand the problem of the indigenes. Undoubtedly, there is a procedural problem in our dealing with Indians and I do not have any reason to hide it."[28]

This misunderstanding between the indigenes and the Otavalo city commission, though incomprehensible to some government authorities, nevertheless was clear to others. The FTP on January 17 determined that the municipal government was responsible "for the criminal acts committed against the indigenes . . . as a result of the feudal system that prevails in a country governed by reactionary elements of the dominant classes."[29]

Undeterred by the uprising and killing of innocent indigenes, as well as impervious to the attacks on them by the national press, Jaramillo and the Asociación 31 de octubre continued their calls for the building of the hotel-casino on the lake. The senator commented,

> My intention was not to wound, nor to do any damage nor to kill any indigene. I was not in any spirit to engage in battle. Sadly, when I arrived at the place where the events were unfolding it was impossible to contain the pitched battle. I have always defended the Indian and I have always been concerned with his/her progress. . . . Painfully it is necessary to continue in the undertaking of building the hotel. Tourism is the only resource that provides economic growth to this region[,] which is dying of hunger and lack of work.[30]

Representatives of the Asociación 31 de octubre echoed Jaramillo's declaration. In the January 15, 1959, edition of the capital newspaper *El Comercio*, they restated their position that Otavalo urgently needed a tourist hotel and that the best location for one was the spot they had already chosen.[31]

Over the course of the next three years a criminal case against Jaramillo would be opened three times. Jaramillo was found innocent each time. In the final case the Supreme Court preempted any new trials, dismissing indefinitely all charges against the senator because of lack of evidence.

The following section examines the misunderstanding that Jaramillo and Bustamante identified as the source of the problems between indigenes and the city council, through the lenses of progress, political ecology, and Andean cosmology.

Modernizing Lake San Pablo

The misunderstanding to which Jaramillo refers in his final analysis of the violence at Lake San Pablo in 1959 was actually the latest in a long line of "misunderstandings" that pitted state actors and political and economic elites against indigenous communities. Throughout most of the twentieth century Otavalo's municipal government and nonindigenous private interests sought control of various parts of the lake. The potential for tourism, however, is what inspired authorities to act unilaterally in the name of progress and development. Each time, however, they encountered resistance from native communities.

The early twentieth century witnessed efforts to "beautify Lake San Pablo" by municipal authorities and individual landholders with the introduction of eucalyptus trees throughout the lake basin. The city also oversaw the introduction of rainbow trout into the lake. Finally, city planners saw the potential for tourism with the creation of a lakeside road. Native peoples all along the shoreline found ways to express their dissatisfaction with these "improvements" through sabotage, theft, and poaching. For the indigenes of Lake San Pablo, the lake held a value that was difficult to quantify. Certainly, access to totora reeds along the shoreline and the imba fish (from which Imbabura Volcano gets its name, as does Imbakucha, the indigenous name for Lake San Pablo) are part of the material culture of indigenous peoples around the lake. However, the lake and adjoining mountain are also central to how Andeans vertically map their ecosystem. More important still, they are essential elements of Andean dualism and sacred space whereby the alignment of gendered *huacas* (sacred sites) produces extremely sacred areas within that alignment. Pucará Bajo de Velásquez falls within the alignment of the female huaca Lake San Pablo and the male huaca Imbabura.[32]

Pucará Bajo de Velásquez sits to the east of *parroquia* Eugenio Espejo and on the northwest shore of Lake San Pablo. It is part of the visual alignment that connects the masculine Reyloma with the feminine Na Sa de Agua Santa and with the feminine lake and masculine Cusin Hills. Moreover, the reverse direction of the alignment connects the masculine Reyloma with the feminine San Juan. Every day the indigenes of Pucará Bajo de Velásquez would have cast their eyes on their beautiful lake in the forefront and the majestic mountain in the background and know that they were standing on supremely sacred lands. This cosmological importance of a sacred visual alignment connecting Pucará Bajo de Velásquez to Lake San Pablo and the Imbabura Volcano would have been inconsequential to Otavalo's commission.

Reexamining the 1959 Massacre

There exists very little information of how the indigenes viewed the municipality, the lakeshore project, or their rights in relation to the lake. However, complaints and petitions before and after the massacre over persisting tensions between Otavalo's city council, the Ministry of Social Welfare (and comunas subministry), and the Pucará Bajo de Velásquez comuna do shed light on relations between whites and indigenes and the relationship of these two groups with Lake San Pablo.

In the final days of November 1958 the indigenous leadership of Pucará Bajo de Velásquez, including Pedro Bautista Velásquez, the president; Segundo Morales Castañeda, the secretary; Francisco Espinosa Morales, treasurer; and Juan Velásquez Cahuasquí, financial trustee, submitted an eight-point complaint to the minister of social welfare and comunas outlining their grievances against the municipal government of Otavalo and questioning the attitude of the ministry for putting their ideas of progress above the general welfare of the indigenous populations. According to this document, Otavalo's city council on April 18, 1956, began its illegal activities with respect to this project when it decided unilaterally to expropriate comuna lands with the objective of building a casino-hotel and a residential village especially for tourists, more specifically, for delegates attending the Eleventh Inter-American Conference. In the complaint Bautista and others refrained from commenting on the Eleventh Inter-American Conference except to say that as far as the communal lands go, no public act should seriously jeopardize the rights of Ecuador's people, especially its indigenous people, to these lands, particularly since one of the objectives of the conference was to address the issue of democracy in Ecuador and the rest of the continent. The document points out that if the project were to continue without addressing the concerns of the indigenous population at the lake, then this action too would be an attack on democracy, especially since more than 130 indigenous families, "who have not much land on which to grow food or to die on, would be negatively affected, while those who get construction contracts and those rich whites who stay in the hotel would benefit."[33] Finally, it called on the ministry to act proactively on behalf of the indigenous peoples and to call a meeting with representatives of the municipal government, of the comuna Pucará Bajo de Velásquez, of the Eleventh Inter-American Conference, and of the Confederation of Ecuadorian Workers, as well as the senators who represented the workers' region. These measures, the letter adds, would be acts of justice and a true defense of democracy and the Ecuadorian people. Finally, it concluded, "we are disposed to defend our land by all legal means possible: we were born on it, on it our ancestors lived, on it our wives and children lived, and we do not want nor will we permit that our lands, which have been stolen from us since the arrival of the Spanish *encomenderos*, be stripped from us again."[34]

In another letter received by the minister of social welfare and comunas, on December 12, 1958, less than a month before the uprising at Lake San Pablo, comuna officials from Pucará Bajo de Velásquez complained bitterly about plans put forth by Otavalo's municipal government, in anticipation of the Eleventh Inter-American Conference, to build a hotel-casino and resort village—*una ciudadela*—atop lands belonging to the comuna. In addition to criticizing the ministry's failure to act in their behalf, comuna representatives questioned how it could allow the municipal government of Otavalo to deal with the issue in any way that it saw fit. For comuna leaders, this behavior demonstrated that the ministry did "not care about the destinies of the 130 peasant families" who were being forced to confront the "great economic interests who planned to profit from the expropriation of our lands." Moreover, it went on to remind the ministry of its responsibility to serve as arbiter in cases of unlawful expropriations of comuna lands and waters and that its decision to put the matter into the hands of the municipal government was in and of itself unlawful and in opposition to the principles of protecting communal rights. Finally, comuna representatives called on the ministry to revoke that resolution giving the municipal government carte blanche in dealing with the indigenes and demanded that the case follow its proper judicial path, and they called on the ministry to preside over a meeting with the respective representatives of the interested parties. In conclusion, the letter pointed out that if the Ministry of Social Welfare responded proactively to their concerns, then it would indeed be fulfilling its legal responsibilities. On the other hand, it forewarned, if the matter were left unattended, they would hold the Ministry of Social Welfare responsible for whatever tragic events might take place but that under no condition would they permit the theft of their lands.[35] Obviously, the warning was not heeded, and the uprising and massacre on January 8 became an awful chapter in Otavalo's history.

Following the tragedy at the lake, Otavalo's municipal government resumed the hotel-pier project along the Lake San Pablo shoreline, and the indigenous leadership continued to defend the comuna's lands. In a letter to the minister of social welfare and comunas dated September 4, 1959, comuna president Bautista reminded the minister that the municipal council of Otavalo was still trying to build a pier and hotel and was "occupying part of our lands, without previously having addressed this problem with us, so that we might find a solution compatible with the interests of the comuna as with the Municipal government of Otavalo." Bautista added that these illegal expropriations had "suppressed our legal rights," as they "reduce the amount of land destined for pasturing and for purposes of a social character of the comuna." Finally, Bautista concluded that the lands in question "were property, resulting from immemorial titles, of the lands *colindantes y ribereñas* [adjoining and riverine] of Lake San Pablo, the same ones that we have been continuing to frequent and maintain for the purpose of *pastos comunes* [communal pasturelands]."[36]

31. Lake San Pablo. Wikimedia Commons.

In response, Otavalo's municipal council rejected the claims by Bautista, arguing that "the petition formulated in the document" by Bautista "had no foundation" since "the pier in construction is located on the same shore of the lake as indicated by the maps that were duly approved." The missive then stated that since 1900 the municipal government of Otavalo had built several piers on the shores of the lake without ever having received any complaints; that the indigenes of Pucará Bajo knew good and well that any charges of misappropriation of lands had been dismissed; and that the areas where the piers had been built were not communal, since the lands in question were occupied by roads, thereby making this beach sandy and not suitable for pasturing. Lastly, Otavalo's interim municipal president added that both the president of the republic (Ponce) and the minister of government had visited the site of the hotel-pier and had reacted very favorably to the project.[37]

The petitions and correspondence presented here provide a glimpse of how whites and indigenes viewed the lake and the status of white-indigenous relations at the time of the 1959 uprising. Elites saw the lake as a resource to be exploited for the purposes of tourism, particularly since Ecuador was to host the Eleventh Inter-American Conference. "Progress," according to Otavalo's city lead-

32. Inti Raymi dancers, Otavalo. Wikimedia Commons.

ers, would benefit all. The indigenes of Pucará Bajo de Velásquez, however, had a much more layered view of the lake. The flora and fauna of the Lake San Pablo basin served the lakeshore inhabitants in ways that were alien to most nonindigenous peoples. For native peoples, access to totora reeds and other vegetation, either for pasturing animals or for other uses, was part of their collective rights as guaranteed by the 1937 Ley de Organización y Régimen de Comunas (Law of Communes). Moreover, centuries of proximity to the lake had established it as an essential element of the people's identity and cosmology. Space was sacred, as was time; grandparents and grandchildren, it was argued, had lived and would continue to live along the lakeshore. These cultural rights, it was argued, were not negotiable, and any violence, regrettable as it might be, would be the responsibility of both the local and national governments.

That Otavalo's municipal government refused to accept Pucará Bajo's decision not to sell these lands even at double the price, coupled with efforts to persuade the Ministry of Social Welfare and Comunas to cede comuna lands in the name of the public good, reveals an elite arrogance; some people considered themselves as being above the law. Native leaders understood perfectly well what municipal leaders were up to and sought national attention by folding their struggle for rights and democracy into a larger one—that which was to be taken up by the Eleventh Inter-American Conference.

Conclusion

Ecuador never did host the Eleventh Inter-American Conference. Problem after problem led to postponement after postponement, and by 1964 there was no longer support for it within the OAS. New forums for discussing Latin American affairs emerged as the region's political atmosphere was increasingly influenced by tensions between the United States and Cuba. Likewise, the hotel-casino complex envisioned by national and local leaders never materialized on the sacred lands held by Pucará Bajo de Velásquez.

Although tragic, the 1959 indigenous uprising and massacre provide an important opportunity for understanding relations between elites and indigenous communities in the mid-twentieth century. Native people around Lake San Pablo most certainly resisted non-indigenous Otavalo concepts of progress, doubting that the corresponding projects would improve their lives. They certainly did view efforts to encroach upon their lands in material terms. Would the city make good and pay what they offered? Would they want even more land? These are but a sampling of questions that native peoples would have asked themselves. However, there is another important lens through which to view the conflict at the lake. Native people's understanding of sacred space emphasized gendered spatial alignments of which Pucará Bajo de Velásquez was part. Lands expropriated by the city or the construction of buildings that might obstruct the alignment might be permissible to individuals, but to the community as a whole it was an attack on its collective patrimony. Upon stating that the conflict at the lake was simply a matter of misunderstanding, Jaramillo and others had it partially right. Whites did not understand the importance of land and water to native peoples; however, indigenes had a very good idea of what these negotiations over land, the building of a tourist hotel, and like manifestations of progress might mean for them.

33. Ecuadorian woman. Wikimedia Commons.

Notes

1. San Félix, *En lo alto grande laguna*, 331.
2. San Félix, *En lo alto grande laguna*, 327–31.
3. Martz, "Ecuador and the Eleventh Inter-American," 307.
4. Peñaherrera, *Informe a la Nación*, 95.
5. The exchange rate for sucres to dollars during the last year of the Velasco presidency was twenty sucres to one U.S. dollar. By late 1957, during the presidency of Camilo Ponce Enríquez, the exchange rate had stabilized at seventeen sucres to one U.S. dollar.
6. Peñaherrera, *Informe a la Nación*, 97–98.
7. Martz, "Ecuador and the Eleventh Inter-American," 308.
8. Martz, "Ecuador and the Eleventh Inter-American," 310.
9. *FRUS*, 1955–57, 7:490.
10. *FRUS*, 1955–57, 7:490.
11. Martz, "Ecuador and the Eleventh Inter-American," 310.
12. Martz, "Ecuador and the Eleventh Inter-American," 310.
13. Villavicencio, *Relaciones interétnicas en Otavalo, Ecuador*, 48.
14. San Félix, *En lo alto grande laguna*, 325.
15. San Félix, *En lo alto grande laguna*, 326.
16. San Félix, *En lo alto grande laguna*, 326.
17. San Félix, *En lo alto grande laguna*, 326.
18. San Félix, *En lo alto grande laguna*, 327.
19. San Félix, *En lo alto grande laguna*, 327.
20. Quoted in San Félix, *En lo alto grande laguna*, 328.
21. San Félix, *En lo alto grande laguna*, 330.
22. San Félix, *En lo alto grande laguna*, 329.
23. La Confederación de Trabajadores del Ecuador, "Comunidado," *La Nación*, January 12, 1959.
24. San Félix, *En lo alto grande laguna*, 329.
25. San Félix, *En lo alto grande laguna*, 331.
26. *Huaca* in this context refers to a piece of pottery that during pre-Columbian burials would have been placed in the tomb. It is not uncommon today for indigenous people to safeguard them in their homes. *Huaca* also refers to sacred spaces, such as mountains, or lakes.
27. San Félix, *En lo alto grande laguna*, 332.
28. Quoted in San Félix, *En lo alto grande laguna*, 331.
29. Quoted in San Félix, *En lo alto grande laguna*, 333.
30. Quoted in San Félix, *En lo alto grande laguna*, 334.
31. Asociación 31 de octubre, "La intervención de la Asociación '31 de octubre' en los proyectados muelle y hotel de San Pablo," *El Comercio*, January 15, 1959.
32. Caillavet, *Etnias del norte*, 415.

33. Comuna Pucará Bajo de Velásquez to Ministro de Comunas y Previsión Social, n.d. (ca. late November 1958), Folder 62, Comunas and Communities, MAG.

34. Comuna Pucará Bajo de Velásquez to Ministro de Comunas y Previsión Social, n.d.

35. Comuna Pucará Bajo de Velásquez to Ministro de Comunas y Previsión Social, December 12, 1958, Folder 62, Comunas and Communities, MAG.

36. Bautista Velásquez to Ministro de Comunas y Previsión Social, September 4, 1959, Folder 62, Comunas and Communities, MAG.

37. Ilustre Concejo Municipal to Ministro de Comunas y Previsión Social, October 15, 1959, Folder 62, Comunas and Communities, MAG.

Bibliography

Archives and Manuscript Materials

Ministry of Agriculture Archive, Quito, Ecuador (MAG).
Ministry of Foreign Relations Archive, Quito, Ecuador (MRE).

Published Works

Caillavet, Chantel. *Etnias del norte: Etnohistoria e historia de Ecuador*. Quito: Abya Yala, 2000.

FRUS (*Foreign Relations of the United States*), 1955–57. Volume 7, *American Republics: Central and South America*, edited by Edith James, N. Stephen Kane, Robert McMahon, Delia Pitts, and John P. Glennon, document 490. Washington DC: Government Printing Office, 1988. https://history.state.gov/historicaldocuments/frus1955-57v07/d490.

Martz, Mary Jeanne Reid. "Ecuador and the Eleventh Inter-American Conference." *Journal of Inter-American Studies* 10, no. 2 (1968): 306–27.

Peñaherrera, Luis Antonio. *Informe a la Nación*. Vol. 2. Quito: Ministerio de Relaciones Exteriores, 1955.

San Félix, Álvaro. *En lo alto grande laguna*. Otavalo: Instituto Otavaleño de Antropología, 1974.

Villavicencio, Gladys. *Relaciones interétnicas en Otavalo, Ecuador: ¿Una nacionalidad india en formación?* México DF: Instituto Indigenista Interamericano, 1973.

THREE

Politics, Projects, and Postwar Possibilities

9

An Alliance for Tourists

The Transformation of Guatemalan Tourism Development, 1935-1982

EVAN WARD

> Now only solitary chicleros cross El Petén,
> The vampire bats nest in the stucco of the friezes,
> The mountain pigs grunt in the evening,
> The jaguar roars among the towers—the towers among the roots—
> The coyote, far away in a plaza, barks at the moon,
> And the Pan American plane flying over the pyramid.
> But will the ancient katunes return one day?
>
> —ERNESTO CARDENAL, "Las ciudades perdidas"

In his poem "Las ciudades perdidas" the Nicaraguan poet-priest Father Ernesto Cardenal juxtaposed the isolated, almost timeless world of Guatemalan rubber tappers with the imminent arrival of throngs of American tourists in the shadow of an unidentified Mayan temple, which might as well have been at Tikal. By the time Cardenal had written the poem in 1960, Pan American Airways' local subsidiary, Aviateca, had been shuttling tourists to the ruins of the Mayan city-state of Tikal for a year. The volume of tourists skyrocketed during ensuing decades, culminating in the arrival of more than two hundred thousand visitors to the site in 2006. While the matchless beauty of Mayan ingenuity accounted for the arrival of such numbers of tourists, growing ties between American institutions shaped the culture of Guatemalan tourism.

Of particular note, Guatemala's government aligned itself strategically with the United States following the Counterrevolution of 1954. Planning, technical assistance, and the Alliance for Progress in particular enabled successive regimes to promote monumental architecture as the premiere local tourist draw during the Cold War, a trend that coincided with Guatemala's efforts to develop its northeastern department—El Petén. Such a strategy mimicked similar efforts by the Mexican and Peruvian governments (of which the Guatemalan government was well aware, as discussed below) to capitalize on an aspect of national heritage that could double as a tool for economic development.

The First Frontier of Guatemalan Tourism Development, 1933-1945

Twentieth-century tourism in Latin America passed through three distinct phases, each distinguished by the nature of relationships between local governments (both national and regional), private capital, and international aid organizations.[1] From the inception of mass tourism, which began in Central America and the Caribbean with the arrival of United Fruit Company vessels in the 1890s, local governments worked directly with domestic and foreign interests to promote tourism development. This constituted the "first frontier" of Latin American tourism development. By the early 1960s international aid organizations, including the United States Agency for International Development (USAID), the United Nations Educational, Scientific, and Cultural Organization (UNESCO), the Organization of American States (OAS), the World Bank, and the United Nations Development Program (UNDP), had put their technical resources behind tourism development and cultural preservation. Assistance from these organizations generally translated into low-interest loans made by the same entities to support tourism development. In the case of Guatemala it was the World Bank, the OAS, the United Nations, and the USAID that promoted such efforts through specialist consultations and loans earmarked for restoration efforts at places such as Tikal. By the mid-1980s, however, such strategies had failed to generate meaningful economic development. To be sure, even amid a civil war, an increasing number of tourists at Tikal generated revenue, but ultimately such earnings did not reach levels sufficient to provide sustainable livelihoods for Guatemalans in the tourist trade or to catapult the nation to higher levels of development. Thereafter in Guatemala, as well as the rest of the emerging world, globally integrated corporations (e.g., cruise lines, hotel chains, and airlines) more successfully promoted free-market mass tourism, while grassroots initiatives championed sustainable tourism projects.

Guatemala's experience began in the early decades of the twentieth century. Strongman and germanophile Jorge Ubico lorded over the country as dictator. Following in the colonial tradition, his government exacted labor in lieu of cash from natives to satisfy tax obligations. This brand of *indigenismo*, or governmental efforts to integrate native communities into a modern economy, also influenced tourism offerings. Each summer the combined forces of private capital and the Guatemalan state attracted thousands of foreign tourists to the country's International Exposition, which featured displays of recent technological advances and the enduring culture of indigenous communities. This curious blend of modernity and tradition piqued the interests of international visitors, who subsequently took tours of Lake Atitlán and Chichicastenango, among other locales in the Mayan Highlands.[2]

The popularity of indigenous village tours benefited from the availability of transportation between the United States and the Central American republic. Easy access to both the Atlantic and Pacific Ocean ports accommodated relatively short travel times from both the Atlantic and Pacific coasts of the United States. On the Pacific coast Grace Line offered transportation from San Francisco to the Guatemalan port of San José. United Fruit Company's Great White Fleet presented a variety of trips—eight, twelve, fifteen, or nineteen days in length—to Puerto Barrios, Guatemala, from the eastern coast of the United States. The United Fruit Company also owned the major railway that connected San José and Puerto Barrios with the Guatemalan capital. Tourists boarded trains headed for the capital, Guatemala City, from ports both east and west. Following a day of acclimation in the capital, visitors piled into cars for a multiday excursion to Antigua (the former colonial capital ruined by earthquake in 1773), the wonders of Lake Atitlán, and finally a visit to Indian markets in towns like Chichicastenango.

A lavish twelve-page brochure, published by the United Fruit Company in the early 1940s, promoted "A Trip to Guatemala by the Great White Fleet." The text featured a glossy overview of the conquest of Guatemala, then made a case for its contemporary appeal, noting, "To an enviable degree she combines the charm of the old world with modern comforts and appointments. She is the veritable Switzerland of the Western World and is making an aggressive bid for the attention of the North American travel market." The brochure then outlined an eight-day tour in the Mayan Highlands. Images of ethnic Maya people accented the itineraries alongside pictures of Antigua and Guatemala City. A man dressed in traditional jacket and headpiece, clutching his red sash as he stands beside a large piece of pottery, greeted the reader on the first page. Leafing through, readers encountered a burnished Maya woman, with a pottery vessel tilted on her head. The following page featured colonial architecture from Antigua, the Temple of Minerva in Guatemala City, and a man encumbered with a bundle of clay vessels and walking away from the photographer. The final pages highlighted a woman weaving at a loom, a marimba player, and men fishing in their traditional attire with the mountains along Lake Atitlán visible in the distant mist. These illustrations emphasized the connection between indigenous communities and the ongoing exploitation of their images for the benefit of national and international commerce.[3]

To be sure, archaeology was not absent among Guatemala's tourist offerings. The United Fruit Company created archaeological parks for the benefit of tourists, as well as the Guatemalan public, but these attractions contributed less to tourism receipts than did trips to native villages to the south and west. The two most prominent locations, situated at opposite ends of the country, were Quiriguá, near Puerto Barrios on the Atlantic coast of Guatemala, and Zaculeu, in the west.[4] Scholars hired to work at these sites had an explicit understanding that their efforts were not merely scientific in nature but that the sites would also serve as draws for tourism. From a business standpoint it does not appear that Quiriguá and Zaculeu contributed significantly to United Fruit Company tourism in Guatemala. A visit to Quiriguá figured as an optional day trip for tourists who had purchased the Guatemalan Highlands tour. This apparent lack of marketability underscores the changes in Guatemalan tourism as development shifted toward excavated ruins.

If heritage tourism was in a nascent stage, no one did more to lay the foundation for its emergence than archaeologist Sylvanus Morley. A precocious Harvard-trained researcher, Morley received a $20,000 grant to conduct excavations at Chichen Itza in Mexico's Yucatán Peninsula. Disposition of the grant was blocked, however, by perceptions of instability in Mexico following its revolution. Never one to while away idle time, Morley surveyed the inscriptions at Mayan sites near Mexico until he had permission from the Carnegie Institution and from the Mexican government to proceed in the Yucatán. (The Mexican government had also been troubled by the efforts of Chichen Itza's owner, F. E. Thompson, to dredge the sacred cenote of it underwater contents.) By the end of his career in Central America, Morley had collaborated on projects at Chichen Itza, Uxmal, Quiriguá, various sites in El Petén (including Tikal), and Copán (Honduras). This work on multiple projects was not unusual. What set him apart was his insatiable desire to unearth archaeological wonders in the most remote posts of the Americas for the public to appreciate.

In El Petén, Morley had the opportunity to study the inscriptions at sites such as Uaxactun and Tikal, but the excavation of such sites—Tikal especially—fell to the University of Pennsylvania. Nevertheless, Morley's magisterial *The Inscriptions of Petén*, published by the Carnegie Institution of Washington in 1938, provided additional evidence of the ways in which United Fruit, the Guatemalan government, and academic institutions collectively established the foundations for later tourism in Central America. In the acknowledgments of *The Inscriptions of Petén* Morley first thanked Guatemala's Ministry of Education "for the official sanctions without which this investigation could not have been [conducted]," then the president of the Sociedad de Geografía e Historia de Guatemala, whose research interests, Morley affirmed, have "always enlisted generous support for the investigations . . . of the Carnegie Institution of Washington."[5] Morley went on to thank the U.S. Department of Agriculture for its "cooperation with the Carnegie Institution and the United Fruit Company . . . [toward] the study of Maya archeology."[6] Finally, Morley proffered his gratitude to the United Fruit Company for its assistance throughout the research for and compilation of the book. His statement read like a who's who of extra-governmental authorities in banana republics, including United Fruit Company's Sam Zemurray and the late Minor Cooper Keith.[7]

In sum, archaeological tourism occupied a minor niche in Guatemalan tourism during the first half of the twentieth century. However, its genesis was due less to large government-sponsored projects and more to relationships forged by governments themselves with private companies, like the United Fruit Company, as well as with the scientific institutions that funded projects such as Morley's work at Quiriguá.

Tourism Development as National Development, 1945–1951

After the unforeseen removal of President Jorge Ubico, Professor Juan José Arévalo's swift—and surprising—ascent to power in 1946 initiated the integration of El Petén into the national fabric and thus the possibility of developing Tikal as an accessible tourist attraction. Arévalo's claims to Guatemala's control over British Honduras (today Belize) turned the country's attention to shoring up national sovereignty over the neighboring territory of Petén. Conscious of the precarious state of the future Belize in negotiations and skirmishes between the Guatemalans and the British, Arévalo feared that unless action was taken, the country might also lose El Petén to its imperial neighbors.

In 1947, under Arévalo's watch, the Inter-American Development Commission hired tourism consultant J. Stanton Robbins to prepare an overview of Guatemala's primary tourist attractions and recommend ways they could be enhanced to lure tourists, especially from the United States.[8] The immediate post–World War II period and Europe's war-torn state offered nations in the Americas an advantage in drawing tourists, a window that would remain open until the late 1950s.

Robbins emphasized that archaeology and indigenous cultures set Guatemala apart. He cautioned planners to be sensitive to the class of travelers that would most appreciate these activities. In order to accommodate students and teachers, for example, Robbins advocated keeping costs down so as to enhance exposure to the entire nation. He noted, "The average traveler is not wealthy and this is particularly true of the traveler with interests—the teacher, archaeologist, student, etc., who can do Guatemala the most good. The long-stay traveler really gets to know the country and remains a friend."[9] Likewise, Robbins urged the government to encourage the establishment of summer and winter language schools for foreign tourists, along with "courses in specialized fields in archeology, art, agriculture, etc." Ultimately, the government should highlight the nation's Mayan past. Cobán and Quiriguá could be exploited as immediate attractions, and Tikal might be further excavated in the future. Ever attentive to the challenges of working in subtropical and tropical landscapes, Robbins believed that to be successful, Tikal's excavation must be accompanied by the construction of a small airfield and guesthouse.[10]

In 1948 officials from the National Tourism Office visited Tikal to assess the possibilities for tourism development there. Antonio López, a local entrepreneur from Flores, wrote that several days later a group of ten officials led by Jacobo Arbenz, minister of defense and architect of the coup against President Ubico, arrived in the jungles of Petén to consider building a makeshift airport at the Tikal site.[11] President Arévalo, in the only reference to the Mayan ruins in his memoirs, recalled, "A commission of Engineers [sic] and archaeologists [went] to Petén, with the usual financial protection of the Government, to commence the cleaning of the hills that covered—until then—the monumental Mayan ruins of Tikal."[12] Several years later the *New York Times* gave six lines in its June 2, 1951, edition to announce the "Air Strip in Guatemalan Jungle." Although brief, the *Times* article noted the dual purpose of the airstrip, as it had been constructed by the Ministry of Defense for purposes of national security but also to "[make] the little known Mayan ruins of Tikal easily accessible to tourists and archeologists."[13] Perhaps as a reward for his efforts in clearing the airstrip, the government awarded Antonio Ortiz the sole right to build a *posada*, or inn, at Tikal.[14] Thus, at the close of President Arévalo's tenure in office, plans for tourism development at Tikal, recommended by the U.S. consulting firm of J. Stanton Robbins, moved ahead with the dual purposes of leisure and strategic defense.

Tourism, Transitions, and Counterrevolution, 1951–1954

The transition from President Arévalo to President-Elect Jacobo Arbenz (who had been the instigator of the 1945 ouster of Ubico) in 1951 has often been portrayed in the historiography of the era as one of continuity. From the standpoint of El Petén's development, however, the two periods were markedly different. In his thorough assessment of presidential contributions to El Petén's economic emergence, Guillermo Pellecer Robles, who headed the Guatemalan agency for developing Petén (FYDEP), noted that Arévalo's government was "focused on the development of Petén," as well as other initiatives there that never materialized "due to a lack of political will." In contrast, Pellecer Robles noted, during Arbenz's presidency, "there was no political incentive anywhere in the department that brought Petén closer to the national economy."[15]

Persistent whispers of undue communist influence plagued President Arbenz's tenure in office. True, the Guatemalan Communist Party participated in the political processes of the republic and had access to President Arbenz, but its actual impact on national policies is debatable. Arbenz nevertheless proceeded with a land reform effort and procurement of a weapons arsenal from Czechoslovakia that drew the ire of the Eisenhower administration. The land reform targeted foreign landholdings, much of which was in the hands of the United Fruit Company, whose controlling interests occupied high positions in the U.S. government. The arrival of Soviet-bloc weapons in Guatemala in 1954 triggered U.S. sanctions against the Central American republic, which was a member of the Organization of American States, and set in motion counterrevolutionary Central Intelligence Agency plans to topple the Arbenz regime. The CIA tapped Carlos Castillo Armas, a lieutenant colonel who had opposed the presidency of Arévalo, as the instrument to lead a coup against the president from a base within Guatemala. With a skeleton army, minimal air support from the United States, and the bishop of Guatemala City's fervent opposition to Arbenz, Castillo Armas toppled the legitimately elected head of state, reversed land reform policies, and aligned Guatemalan foreign policy with that of the United States.

Tourism played a minor, though not inconsequential, role in the CIA's ouster of Arbenz in 1954. During the previous half century Guatemalan tourism had been a darling of the travel-writing world. In the months before the coup journalists painted a dire picture of Guatemala. Sydney Gruson captured the mood of tourists in Guatemala City, writing in the *New York Times*, "The crisis has left the capital a city of little gaiety. The trickle of tourists who came intermittently during the past few years has virtually stopped. Hotels lobbies, restaurants, and night clubs are empty, melancholy places. The business depression has worsened as people hoard their assets against an unknown future."[16] In September 1954, however, Julia Batres painted a different picture, extolling personal safety under the new Castillo Armas regime.[17] Not only were Guatemala's traditional tourist attractions more appealing than ever but new tourism linkages between the new administration and the United States had been extended with the opening of the Guatemala Tourism Commission office in New York City. Two months later the *Chicago Tribune* published a similar article, written by Jules Dubois. Entitled "Guatemala Opened to Winter Travel: Boom Foreseen," Dubois noted rather facilely, "It will be like old times as the land of the Quetzal bird is expected to have a boom season now that the Communists have been booted out."[18]

The Castillo Armas administration accelerated the pace for tourism plans in and around Tikal. In its November 25, 1955, issue the journal *Science* announced that the museum of the University of Pennsylvania had signed a contract with the Castillo Armas gov-

34. Tikal airfield, 1971. Wikipedia Commons.

ernment to begin excavations at Tikal. While archaeological agreements between governments and prestigious universities typically facilitated excavation of premiere archaeological ensembles, this agreement contained a unique directive: the university's scientists would work to stimulate tourism while they worked to preserve the ruins. As Edwin M. Shook, director of the University of Pennsylvania's archaeological team, noted in his first report from Tikal, his staff had built roads to accommodate the steady increase in numbers of visitors arriving at the site. Furthermore, the Tikal airstrip facilitated the arrival of even more tourists. This spurred on the American team to excavate as many of the ancient structures as they could.[19] Likewise, in reflecting on recent excavations, Shook emphasized the two-pronged approach: to excavate key structures at the site and prepare the complex for mass tourism.[20]

The contract, drawn up officially between Shook, as representative of the University of Pennsylvania Museum, and Enrique Quiñones, minister of public education of Guatemala, gave exclusive rights to the University of Pennsylvania to excavate at Tikal, as well as any other archaeological sites in El Petén, for the period of five years beginning January 1, 1957. The agreement required the university to expend at least $20,000 (U.S. dollars) each year, with the possibility of relinquishing the exclusive license if it did not. The contract also specified that Guatemalan workers and U.S. scientists alike should preserve the integrity of monumental structures. Where restoration was needed, the government enjoined the university to use its financial resources to prevent the further deterioration of structural features. The university museum's scientists would deliver recovered objects to the Guatemalan Institute of Anthropology and History for display in Guatemala or request an exemption to take pieces out of the country to Philadelphia for research for a period of up to five years. In financial terms the agreement provided the university with an exemption from all taxes and duties related to research-related materials.[21]

The historic agreement acknowledged the precarious state of transit between Guatemala City and El Petén. Logistically, carrying out Tikal's restoration proved to be nearly impossible when approached from the ground. Airlifts of supplies, machinery, and people were the only viable avenue of transport in the late 1950s. Hoping to leverage the maximum amount of specialized work from the funds provided by the University of Pennsylvania, the Guatemalan government included a novel stipulation in the contract that it would provide free air transport between Guatemala City and ports adjacent to Tikal to the university's team of scientists. In a nod to the dual mission of the expedition, the government also agreed to maintain the airstrip at Tikal, where regularly scheduled Aviateca flights commenced in 1959.

Castillo Armas's personal visits to Tikal further illustrate his interest in what would be named the nation's first national park. Presidential junkets to Tikal began as early as the first season at the site, in 1957. While Shook was preoccupied with finding potable water beneath the forest floor, President Castillo Armas and a few select friends invaded the camp from time to time. The first presidential contingent, comprising seven airplanes and an undetermined number of people, made an aerial procession onto the fifteen-hundred-meter dirt airfield. The visitors then walked to the famed Acropolis to eat lunch. There the VIPs came face to face with the untamed forest. Recalled Shook, "We were just passing the Temple of Inscriptions, which was covered by jungle and bush, when we heard horrendous screams from the ladies. We looked back and caught a glimpse of a huge puma that had landed about half way down the pyramid in one bound. On the next bound, the puma landed at the base of the pyramid, at the feet of the ladies, and then it disappeared into the jungle like a streak. . . . There was chaos for about ten minutes."[22]

If the tours by dignitaries often annoyed Shook, he also found their naïveté amusing and chalked the experiences up as part of the university's mission in preparing the park for tourists. Speaking of Castillo Armas's initial visit to the park, Shook recalled, "Not that they knew anything about its history, but those temples are impressive to anyone. And I've always felt that the rainforest itself, with all the animal and bird life, is just about as exciting as the archeological ruins." Apparently, the appeal had staying power. Shook added, in retrospect, that "presidential parties came at least once a season and sometimes twice."[23] The next year Castillo Armas returned to Tikal dressed in army regalia. A photograph in Shook's memoir depicts the president walking around carved stelae while Shook explains its markings to the president and his company.

Castillo Armas's assassination in 1959 did not stem the wave of media, diplomats, socialites, and politicians who called on Ed Shook and his successors. Shook's memoirs contain a picture of Castillo Armas's successor, President Miguel Ydigoras Fuentes, visiting the ruins. For a director who was ambivalent toward the growing trickle of tourists, Shook readily acceded to an agreement that the museum made with *Life* magazine to allow photographers the run of Tikal's ruins for an exclusive montage. *Life* writers and photographers arrived on April 2, 1958, and a frenzied exposure of film followed. These scenes moved Shook to reflect on his own life as a tourist in Guatemala. "Never have I seen so much film exposed, both in color and in black and white," he exclaimed. "Now I know how the Maya feel in Chichicastenango!"[24] Two weeks later, on April 15, 1958, U.S. ambassador Lester Mallory, President Ydigoras Fuentes, and a contingent of fifty individuals arrived at Tikal in three airplanes. Shook recalled that he drove the new president and his entourage to the site and then marveled at the distinguished

35. Tikal temple 1. Wikimedia Commons.

photographers among them, including Fritz Goro from *Life* magazine.[25] By 1959 the Guatemalan government had "elaborate plans for the development of the National Park this year."[26] These plans coincided with the advent of regularly scheduled air service from Guatemala City on Aviateca Airlines. Service ran three days a week, with special excursions on Sundays.

An Alliance for Progress, an Alliance for Tourists

In an indirect way the Cuban Revolution transformed Tikal because its momentum set in motion increased U.S. aid throughout the hemisphere, including through the Alliance for Progress. The aid program, announced one month before the ill-fated 1961 Bay of Pigs invasion, proposed modest though comprehensive political and economic change throughout Latin America. Alliance administrators identified tourism as a key area that could simultaneously increase economic growth and jump-start development throughout the region. While grants and loans would be made to individual countries, the structure of funding anticipated a coordinated approach to "sea, sun, and sand" and heritage tourism throughout Central America, including in Guatemala.

By 1965 USAID had created the financial infrastructure for Latin America to pursue regional tourism development. Members from each nation of the new Central American Common Market met in January of that year and began touring the Central American states to assess attractions that would make the region more competitive in the global tourism market. The recommendations of the consultation team, Porter International Company (PIC), selected by the Central American Bank for Economic Integration to draw up a report on their findings, echoed many of the directives given by J. Stanton Robbins nearly two decades earlier. PIC continued to stress decentralization, which would be facilitated by roads planned for immediate completion, including the Pan-American Highway. "At such time the travel map of Central America will be radically different than it is today," the consultants rosily prognosticated. "A motorist, having reached Guatemala City, will not be restricted to the Pacific or Pan American Highways, but will be offered a variety of attractive alternatives. Air travelers, instead of proceeding on after a few days, as is now so often the case, will be tempted to switch to a car and tour, as time allows, Esquipulas, Copán, Lake Izabel, and the Caribbean."[27]

The PIC report also spent considerable time weighing the advantages of accelerated development at Tikal. Although Petén itself was in dire need of roads to connect it to Guatemala City, Alliance for Progress planners believed that air access to the site adequately accommodated interested tourists. This may have been more a cost consideration than a feasible alternative to greater national integration, but it also reflected the experience of privileged tourists, including Guatemala's presidents, during the previous decade. One of the objectives of the Alliance for Progress had been to spark private investment (for every two dollars contributed by the United States, Latin American nations were expected to provide eight, either through public or private channels). Thus, both the monumental appeal of the site as well as the staggering cost to restore the lost city required special treatment. As the report observed, "As time passes and Tikal is gradually freed from the jungle, it becomes even more apparent that this is a site every tourist should have the opportunity to see. Related to this is the more fundamental position that the national value of Tikal transcends other considerations. We are led to the conclusion that all phases of Tikal's development should be a national responsibility."[28] The question remained, however, as to whether or not tourism represented the "magic bullet" to boost development of activities outside the service sector.

Two years later, in 1967, the United Nations dispatched Dr. J. G. Ramaker to Central America to prepare a follow-up report to the Porter study for the Office of Technical Cooperation. He championed further development at Tikal, including construction of a new hotel and improved roads, but made peripheral observations that likely spoke to the weaknesses of the Alliance for Progress in Central America. In the cover letter to his report he noted that in "many cases technical assistance is needed; I noted it in several sections of my report. This proves to be a necessity, often where the knowledge to act is available, *but where pressure from outside seems a condition to get things started.*"[29] Ramaker did not elaborate on the sources of outside influence, but it is quite possible that it came from Alliance for Progress and USAID officials stationed in the region. Proposing programs and cashing checks for project completion were one matter; cooperation from hosting governments was entirely different. Years later John T. Bennett, who had served as chargé d'affaires in Guatemala City during the late 1970s (and became an ardent admirer of Tikal), observed that USAID specialists tried to implement potentially worthwhile projects but were "limited by the ineffective and corrupt government and more importantly, the right wing, particularly the old elite, who didn't want Americans messing about with their power structure."[30]

Nevertheless, Tikal represented an exceptional opportunity for tourism development. Therefore, the Guatemalan government contracted with the U.S. National Park Service in 1971 to design a strategy for Tikal's development. The strategy document was entitled "Master Plan for the Protection and Use of the Tikal National

Park." A new airport at Flores topped the list of improvements envisioned in the master plan. As the planners noted, "the construction of an international airport is an important pre-requisite to accelerate the growth of tourism in the region."[31] Domestic political concerns, namely a fear that peasant insurgent groups might spiral out of control, may have dovetailed with touristic purposes of the new structure, for "[this] site was chosen . . . for construction of the international airport after consulting with Colonel Mendoza, Director General of Civil Aviation in Guatemala, and with authorities of Aviateca Airlines." The airport would accommodate Boeing 727s, as well as house air force jets. "As soon as the new airport is completed," the report stated, "air traffic would be significantly increased with direct flights from Honduras, Belize, various points in the United States. Then El Petén . . . can be incorporated within the tourist packages with air access to archeological sites in Mexico and Central America."[32]

The strategic plan also proposed expanded hotels, dormitories, and campsites near the park. Antonio Ortiz's private lodge, Posada de la Selva, already had an expansion project under way that would double the number of tourist cabins, from thirty-two to sixty-four. Building on a trend that catered to young domestic travelers, a trend known throughout Latin America as "social tourism," the planners proposed construction of dormitories that would house students visiting the park for educational and recreational purposes. The proposed dormitories were admittedly a "low priority" in the overall master plan, and, if social tourism did not catch on, "the space . . . could be used as an area of expansion for camping and for hotel cabins, or as an open space between them."[33]

Regardless of its shortcomings, the National Park Service's master plan provided a blueprint for development over the next twenty-five years. Its more immediate purpose, however—that of convincing the Alliance for Progress to release funds for construction of the airport, road, and restoration projects—left much to be desired. The United States's Regional Office for Central American Projects rejected its findings and asked for a more sophisticated assessment. It would not be until 1974, when the Guatemalan government, in concert with the OAS and USAID, put forth an acceptable plan, that loans were forthcoming.[34] In addition to the scheduled additions at Tikal, projects in Honduras and El Salvador received assistance. A $15 million loan earmarked for these purposes underwrote projects intended to boost regional tourism receipts.

In the meantime the OAS commissioned engineer Pablo Leclerq to travel to Guatemala during April and May 1973 to work with two other OAS officials on preparing a long-term strategy for Guatemalan tourism. While the National Park Service's proposal offered a viable model for enhancing the capacity of Tikal to receive more tourists, the question of tourism's ability to promote development still lingered. It was LeClerq's study that offered the sobering response to such a pressing question. The report bluntly calculated the futility of state-funded strategies for integrating tourism development into a national economic plan—whether that plan aimed for growth, development, or both. Nevertheless, it concluded that government participation in the short term was the only avenue to capturing tourists flocking to Mexico and the Caribbean. The new strategy tilted Guatemalan tourism schemes toward a model that capitalized on the niche role of Tikal as a catalyst for sea and sun beach development at Izabel, Guatemala.

A small number of interviews conducted with tourists who had visited Tikal, as well as other cultural tourism sites throughout the Americas, including Machu Picchu and Chichen Itza, suggested that Guatemalans could harvest "low-hanging fruit" in the form of already established tourist streams, should the right infrastructure be put in place to bring the tourists to the jungles of Petén. The team interviewed 380 visitors to Tikal, of whom 54 percent had also visited at least one of the following archaeological sites: Teotihuacán, Uxmal, or Chichen Itza (all in Mexico); Copán in Honduras; or Machu Picchu in Peru. These tourists were then asked to compare their impressions of Tikal to comparable archaeological complexes. From this handful of tourists the consultants concluded that, "save for Machu Picchu, where the preference vis-à-vis Tikal was two to one [(16 versus 8 in zero-sum comparisons)], tourists unanimously expressed a preference for Tikal."[35]

There was no question that Tikal ranked among the hemisphere's superlative archaeological assemblages. More elaborate plans for linking development at Tikal to beach resorts at Izabel, Guatemala, would have to wait until the more elementary stages of airport construction and direct access from Flores to Tikal by road were completed. With the Tikal project analysis in hand, USAID put its imprimatur on the $15 million loan to the Central American states by way of the Central American Bank for Economic Integration, with $2.5 million set aside for the Tikal project. For two years of archaeological restoration Guatemala would get $500,000. Provisions were also included for roads, parking, and drainage improvements, as well as technical upgrades for the new airport at Flores.[36]

In December 1977 work on the expanded project still limped along. "Despite the growing number of tourists," a World Bank report noted, "Tikal's accommodation and visitor facilities are totally inadequate."[37] The new airport opened in 1982 with a civilian terminal and facilities for military aircraft. Progress remained slow, however. Commuter planes continued to carry small numbers of tourists to Flores but with greater frequency. Approximately fifteen thousand tourists arrived at the national park in 1981, a number that increased steadily to more than sixty thousand by the late 1980s. This is surprising, given that the country was wracked by civil war; however, little of the conflict spilled over into Guatemala's northeastern reaches. Thus, Flores's military/civilian airport did not figure strategically into the ongoing war.

It would not be until 1988 that the first Boeing 727 landed at the terminal. Exponential growth in numbers of international tourists visiting the park ensued, reaching a peak of nearly 250,000 in 2006. However, Guatemalan visitors to the park steadily increased as well, suggesting that improved roads between Guatemala City and Flores made as much of a difference in stimulating attendance as the new airport did.

Conclusion

Paradoxically, neocolonialism has played a prominent role in the history of Guatemala since its independence from Spain (1821) and

Mexico (1824). Within a generation of its liberation, entrepreneurs and swashbuckling ne'er-do-wells had descended on Central American republics to monopolize the land and the labor of its indigenous inhabitants. In the mid-twentieth century Tikal's evolution illustrates the persistence of imperial influence through the lens of tourism development, despite tourism's limited efficacy as a sustainable growth engine. During Guatemala's "Spring of Democracy" (1946–54), the Arévalo and Arbenz administrations laid the groundwork for heritage tourism in El Petén, including at Tikal. Prior to that time government efforts to attract tourists centered on national expositions and auto tours of indigenous villages. In the aftermath of the 1954 counterrevolution that brought Carlos Castillo Armas to power, however, the University of Pennsylvania inked an agreement with the Guatemalan government to excavate selected areas at Tikal to promote tourism and to recover the nation's cultural patrimony. Subsequently, U.S.-affiliated aid organizations offered technical assistance to enhance Guatemalan tourism infrastructure but often did so without considering the scarcity of natural resources, structural limitations of the economy, or the impact of watershed events that could significantly alter expectations. These factors notwithstanding, the Alliance for Progress, spurred on by USAID, made available $15 million for expanded tourism in Central America, a portion of which benefited the national park at Tikal. The nexus between a foreign government, international aid organizations, and the Guatemalan government epitomized the neocolonial linkages that have characterized mass tourism development throughout the modern world.

Notes

Epigraph: author's translation of Ernesto Cardenal's "Las Ciudades Perdidas," Spanish-language text accessed 29 August 2018 at http://www.revistadelauniversidad.unam.mx/ojs_rum/index.php/rum/article/view/7701/8939.

1. See Ward, "Footprints, Frontiers, and Empire," 9.
2. See Little, "Visual Political Economy."
3. United Fruit Company, "Trip to Guatemala."
4. For background on the sites, see Boggs, *Guide to the Ruins of Zaculeu*. On Quiriguá, see Popenoe, *Quiriguá*.

5. Morley, *Inscriptions of Petén*, vi.
6. Morley, *Inscriptions of Petén*, vi
7. Morley, *Inscriptions of Petén*, viii.
8. Robbins, "Preliminary Report."
9. Robbins, "Preliminary Report," 7.
10. Robbins, "Preliminary Report," 16.
11. Ortiz Contreras, "Hotel Posada de la Selva," 18.
12. Arévalo, *Despacho presidencial*, 392.
13. "Air Strip in Guatemala Jungle," *New York Times*, June 2, 1951, 11, accessed in ProQuest Newsstand database.
14. Ortiz Contreras, "Hotel Posada de la Selva," 16–23.
15. Pellecer Robles, *Petén*, 32.
16. Sydney Grunson, "Guatemala Grim as Tension Rises: People Look for a Climax to End Crisis—Rumors Add to Mood of Nervousness," *New York Times*, 30 May 1954, 16, accessed in ProQuest Newsstand database.
17. Grunson, "Guatemala Grim," 16.
18. Jules Dubois, "Guatemala Reopened to Winter Travel: Boom Foreseen; Hotels Prepare for Big Increase with Reds Out[,] Look for Big Resort Year in Guatemala," *Chicago Daily Tribune*, November 14, 1954, i1, accessed in ProQuest Newsstand database.
19. Shook, "Tikal Report No. 1," 7.
20. Shook, "Tikal Report No. 1," 18–19.
21. Shook and Quiñones, "Agreement."
22. Shook, "Tikal Report No. 1," 137.
23. Shook, "Tikal Report No. 1," 138.
24. Shook, "Tikal: Problems," 16.
25. Shook, "Tikal: Problems," 16.
26. Shook, "Tikal: Problems," 19.
27. Porter International Company, "The Role of Tourism," 59–60, TE 322/1 LA (140–44), UNTAM.
28. Porter International Company, "Role of Tourism," 64.
29. J. G. Ramaker to Mr. Dabezies, August 30, 1967, cover letter for "Tourism Development in Central America (Including Panama)," August 1967, TE 322/1 LA (140–44), UNTAM (emphasis added).
30. John T. Bennett, interview by Charles Stuart Kennedy, October 2, 1987, Foreign Affairs Oral History Project, Association for Diplomatic Studies and Training, http://www.adst.org/OH%20TOCs/Bennett,%20John%20T.%20and%20Thomas%20Stern.toc.pdf.
31. U.S. National Park Service, "Master Plan," 32.
32. U.S. National Park Service, "Master Plan," 33.
33. U.S. National Park Service, "Master Plan," 76.
34. See Secretaría General del Consejo Nacional de Planificación Económica, "Tikal."
35. U.S. National Park Service, "Master Plan," 63.

36. U.S. Department of State, "Capital Assistance Paper," 70–71.

37. World Bank, "Guatemala: Tourism Sector Review and Project Identification," December 7, 1977, 15, http://documents.worldbank.org/curated/en/319011468246599615/Guatemala-Tourism-sector-review-and-project-identification.

Bibliography

Archival Sources

United Nations Technical Assistance Mission in Latin America, Tourism Development (Organization of Central American States), United Nations Archive, New York City (UNTAM).

Published Works

Arévalo, Juan José. *Despacho presidencial: Obra póstuma*. Guatemala City: Editorial Oscar de León Palacios, 1998.

Boggs, Stanley H. *Guide to the Ruins of Zaculeu, Dept. of Huehuetentango [Huehuetenango] Guatemala: Restoration by United Fruit Company*. Boston: United Fruit Company, 1946.

Little, Walter E. "A Visual Political Economy of Maya Representations in Guatemala." *Ethnohistory* 55, no. 4 (2008): 633–63.

Morley, Sylvanus Griswold. *The Inscriptions of Petén*. Vol. 1. Washington DC: Carnegie Institution, 1938.

Ortiz Contreras, Antonio. "Hotel Posada de la Selva." In *Guatemala: Antecedentes historicos del turismo en Guatemala*, 3:16–23. Guatemala City: Instituto Guatemalteco de Turismo, September 1995.

Pellecer Robles, Guillermo. *Petén, FYDEP, y yo*. Guatemala, 2010.

Popenoe, Wilson Popenoe. *Quiriguá, an Ancient City of the Mayas*. Boston: United Fruit Company, 1927.

Porter International Company. "The Role of Tourism in Central America." Washington DC, 1965.

Robbins, J. Stanton. "Preliminary Report on Tourist Development in Guatemala." Inter-American Development Commission, Washington DC, 1947.

Secretaría General del Consejo Nacional de Planificación Económica. "Tikal: Proyecto de desarrollo turístico." Guatemala, February 1974.

Shook, Edwin M. "Tikal: Problems of a Field Director." *Expedition* 4, no. 2 (1962): 16. https://www.penn.museum/sites/expedition/tikal-problems-of-a-field-director/.

———. "Tikal Report No. 1: Field Director's Report; The 1956 and 1957 Seasons." In *Tikal Reports: Numbers 1–11*. University Museum Monograph 64, facsimile reissue of original reports, published 1958–61. Philadelphia: University of Pennsylvania, the University Museum, 1986.

Shook, Edwin, and Enrique Quiñones. "Agreement between the Guatemalan Government and the University of Pennsylvania, 1957." University of Pennsylvania Museum of Archaeology and Anthropology, Philadelphia.

United Fruit Company. "A Trip to Guatemala by the Great White Fleet." N.d. Copy in author's possession.

U.S. Department of State (Agency for International Development). "Capital Assistance Paper: CABEI; Tourism Infrastructure Loan." February 8, 1973. Copy in author's possession.

U.S. National Park Service. "Master Plan for the Protection and Use of Tikal National Park." 1971. Scanned copy in author's possession.

Ward, Evan R. "Footprints, Frontiers, and Empire: Tourism Development in Latin America, 1840–1959." *History Compass* 12, no. 1 (2014). https://onlinelibrary.wiley.com/doi/abs/10.1111/hic3.12125.

10

"Created by God" (or Columbus?) for Tourism

Building Tourism Fantasy in the Dominican Republic, 1966-1978

ELIZABETH MANLEY

Despite modest efforts by dictator Rafael Trujillo (1930–61) to develop tourism in the Dominican Republic during the 1950s, including the building of several luxury hotels and the hosting of a lavish world's fair, the *industria sin chimeneas* (industry without smokestacks) did not get off the ground until well after his assassination and the subsequent U.S. occupation in 1965. In the late 1960s, however, Trujillo's successor, Joaquín Balaguer, began investment in the industry that had proven itself successful elsewhere in the Caribbean. The Balaguer regime supported tourism heavily, eventually focusing on the development of coastal resort areas like the northern beaches of Puerto Plata and the eastern tip of Punta Cana. Growth continued at a rapid pace through the 1970s and into the 1980s, the period during which the now nearubiquitous all-inclusive resorts appeared on the landscape.

For regime officials the possibilities to create a tropical playground to the great benefit of the Dominican economy seemed massive; as noted by Balaguer's first director of the Dominican Republic's National Tourism Bureau, Ángel Miolán, the country was "created by God for tourism, I have dared say sometimes, excited by the beauty of its potential."[1] Under the Balaguer regime, which would last from 1966 through 1978 and was commonly referred to as the *doce años* (twelve years), the nation consolidated its tourism agenda, solidifying the National Tourism Bureau as the central administrative arm of the industry, creating a tourism police force, encouraging massive foreign investment, and concentrating efforts within the *zonas turísticas* identified by several foreign consultant reports. In the process of this twelve-year effort regime officials laid the rhetorical groundwork for the tourism industry, particularly its theoretical premise that the country was uniquely situated—even fated—to be the region's central hub of travel. By 1994 the Dominican Republic was the most popular destination for tourists in the Caribbean and the island had been imagined as an all-consumable paradise for visitors from the Global North.

The presence and intervention of God—as in having designated an island or islands ideal for tourism or being represented by earthly (tourism) agents—is a recurring theme in the industry literature from the 1960s and 1970s across the Caribbean. In the Dominican Republic, however, Christopher Columbus specifically played a predominant role in that mythologizing, and such evocations highlight the envisioned destiny for the nation. Concurrent with that focus on the "first tourist" was a concerted effort to highlight the country's "natural" beauty, specifically its women. In focusing on the legacies of Columbus, the Balaguer regime and its tourism officials highlighted the many "firsts" that established the island as Spain's first colonial stake in the Americas, as well as the beautiful Dominican women serving as ambassadors in this process of exploration. The government thus constructed a viable fiction of the nation as "discoverable" by travelers, enticingly hospitable for foreigners, and destined to become the premier Caribbean travel destination. The image-making component of this endeavor sought to set the Dominican Republic apart from its regional competition and laid the foundations for the broader industry to rely on these convenient fictions to create an illusory vision of the ideal vacation. In other words, by romanticizing Columbus, sketching out an idealized vacation of "discovery," and presenting Dominican women as the ideal ambassadors to travelers, industry players from the government to the private sector established patterns of marketing that continue into the present and construct the country as a tourism fantasy.

Tourism, Discovery, and Fantasy in the Caribbean

Building a number of campaign promises around modernity and development for the Dominican state, Joaquín Balaguer entered the presidency in 1966 needing to demonstrate, at least on the surface, that the nation was prepared to transcend many of the "backward" ways of dictatorship. Obscuring the fact that he himself had been a key advisor for the entirety of the thirty-one-year Trujillo regime, Balaguer sought to prove to the world that the Dominican Republic was ready to progress beyond the authoritarian tendencies of his predecessor. One of the central components of his development plan was a significant rhetorical investment in the possibilities of tourism. He quickly initiated efforts to reinvigorate the Dirección General de Turismo (General Tourism Bureau), initially formed under Trujillo, and appointed Ángel Miolán as its director.[2] The agency's early, relatively modest plans suggest confidence at the highest levels of the administration and a vision that would ultimately carry the Dominican Republic to its place as a Caribbean tourism hub.

Tourism was of course not a new phenomenon in the Caribbean. Beginning with travel on fruit steamers in the mid-1800s, individuals had been visiting the region in search of relaxation, improved health, good weather, and the exotic for more than a century. Barbados and Bermuda had led the way, with Jamaica, Puerto Rico, and Haiti having become popular destinations by the early 1900s. Much of the early twentieth-century development owed its increased numbers to the fruit trade itself, and "paradise was rediscovered" precisely because of the banana and the steamer ships hauling tropical fruit to distant ports.[3] Across the Caribbean visitor numbers boomed in the interwar years, with Cuba in particular becoming an extremely popular place for U.S. visitors. Tourism surged in the 1950s with significant advances in airplane travel.[4] By the early 1960s the Caribbean had become a popular destination for North American and western European travelers, even though the largest tourism hub—Cuba—was no longer an option for many of them. Dominican officials sought to capitalize on and expand the available market by highlighting the accessibility and beauty of the eastern half of Hispaniola.

This possible paradise in the Caribbean was less created than re-presented to the twentieth-century tourist or potential tourist. From early on, the islands were described to the Old World as a fertile, fecund paradise ripe for consuming, and such imagery became central again in the late nineteenth century as government tourism officials sought ways to promote their islands to potential visitors.[5] Throughout the first half of the twentieth century these images built—together with strategic ad copy and marketing campaigns of various stripes—an Edenic ideal. While the Dominican Republic lagged behind other Caribbean nations in tourism development, tourism officials began their work in the 1960s with plenty of raw material with which to present their nation as the most paradisaical of all, as well as to demonstrate how ready it was for new and modern forms of rediscovery.

As travel to the Caribbean became an achievable reality for a larger portion of the U.S. and western European population, travelers were encouraged to uncover this Eden for themselves. This "rediscovery" or more precisely the repackaging of paradise during the second half of the twentieth century takes on important valences on the island "first discovered" and "most loved" by Columbus. The focus on Columbus by tourism officials—as though the world had forgotten the lush tropical island he once claimed for Spain—provides an important framing for the growth of the tourism industry in the Dominican Republic. The very notion of discovery—finding that perfect island getaway—dominated the islands' tourism campaigns from their beginnings and continued concurrent with the broader construction of the islands as exotic and otherworldly, as well as the specific vision of the Dominican Republic as divinely fated for tourism development.[6] Thus, in the process of individual discovery travelers could create their own tourism fantasy.

This vision allowed visitors to imagine stepping, as explorers, into a place that was completely foreign to their own reality, while feeling welcomed and relaxed. As many tourism scholars argue, the creation of a Caribbean paradise is grounded in keeping the complicated histories of enslavement, subjugation, and colonization present but glossed, much in the same way that "discovery" relies on a romanticized and flattened vision of Columbus and the Spanish conquistadors.[7] In the Dominican Republic the promotion of paradise entailed isolating a set of Columbus-related events in the 1490s, through which government and private-sector tourism promoters not only set up a particular and comfortable "field of power" about that past as a romantic one but also worked new stories of travel into the old.[8] As present-day explorers, tourists would be able to "taste and feel" the "soul of the past" while also not finding themselves burdened by challenging histories or placed too far from modern conveniences.

Caribbean paradise, early tourism promoters were eager to explain, was available for individuals to reconstruct as their own tourism fantasy. In addition to the thrill of maverick exploration, beautiful women awaited to serve as ambassadors for their Edenic escape. Caribbean travel was to become an entirely consumable affair. The construction—and eventual domination—in the Caribbean of the all-inclusive resort built during this period had ushered in the fulfillment of this fantasy by the 1980s. However, in the early years of building the industry, projecting the island getaway as a process of discovery and as an escape from the rigors of daily life accompanied by beautiful exotic guides formed the foundation of Dominican tourism development. It was a carefully constructed conceit built around a flattened colonial narrative and an exotification of Dominican female beauty that sought to sell the island as the perfect old-meets-new escape.

From Primitive to Paradise, 1966–1969

During the first several years of the Balaguer regime, newly appointed officials in the tourism bureau sought to demonstrate the viability of the industry to other government functionaries, a global audience of potential tourists, and Dominicans themselves. It was a challenging task, given the paucity of existing resources and widespread skepticism among the Dominican population. A popular rhyming refrain directly questioned the tourism director's logic and efforts: "¿Y los turistas donde están? En la cabeza de Miolán" (And the tourists, where are they? In Miolán's head). Yet officials believed tourism to be "a source of capital generation for the general development of the national economy" and a demonstration of Dominican modernity and progress.⁹ Within the first few years tourist numbers expanded considerably, rising from twenty-eight thousand to forty-five thousand in the first year, then to seventy-four thousand annually in 1969, and facilities grew along with those numbers.¹⁰ Seeking to prove that tourism advancement was crucial to broader national economic growth and stability, tourism bureau officials focused on how the industry fit into larger national plans and what could be done with minimal to no investment. They highlighted the need to move away from what they saw as a primitive, nontourism-focused model of growth to one that showcased the nation's potential as a tropical fantasy for travelers from the Global North.¹¹

Plans for development emanating from the National Tourism Bureau in the early years were relatively inchoate but centered on the natural beauty (and beauties) of the Dominican Republic, its colonial heritage, and the vast swaths of land available for foreigners to discover and purchase. In a 1967 report produced by José D. Vicini, an estimated forty thousand plots of five thousand square meters of uncultivated land could be purchased and then auctioned off to the highest U.S. bidders looking for a "new paradise," resulting in a potential windfall for the industry of just under $500 million.¹² Such investors, according to Vicini, were searching for "picturesque landscapes" and hoping that "these sunny isles are relatively unndiscovered."¹³ As Vicini—heir to a sugar empire—and others argued, the tourism project was largely about creating a discoverable paradise that to this point was not yet known to Global North investors.

Tourism officials were well aware that the eastern side of Hispaniola, unlike their neighbor Haiti, received virtually no tourist attention and was ripe for discovery by investors. Bemoaning the absence of the country from the *New York Times* supplement "Caribbean Showcase," José Andrés Aybar Castellanos, who was head of the National Development Commission (Comisión Nacional de Desarrollo, or CND) noted that although "there [was] very effective advertising about the islands around Santo Domingo," there was nothing written "with respect to our country, save two ads from private companies, and [they were] very shallow."[14] Such reflections were not off the mark. For example, *Caribbean Beachcomber*, a magazine that purported to represent the entire region and was published by several Caribbean tourism organizations, provided minimal coverage of the country in the late 1960s. In their "Caribbean Vacation Planner" the Dominican Republic warranted only a quarter column.[15] Noting a "long and complicated" political history that served as a discreet warning, the column pointed potential travelers toward the capital only, with its colonial monuments and the beautiful nearby beaches of Boca Chica. As noted by a Pan American World Airways official in 1968, there remained much work "to combat certain unfortunate impressions that may reside in the minds of American tourists."[16]

Endeavoring to challenge these "unfortunate impressions," tourism bureau officials found allies in the burgeoning tourism press. The magazine *Bohío Dominicano*, begun in winter 1966, was generally supportive of the work of the National Tourism Bureau and assisted in the project of making the country discoverable to potential tourists. Published quarterly, *Bohío* was the first regular tourist publication and was marketed directly to foreign visitors. It was conceived and run by Luis Caminero and his wife, Rita Cabrer de Caminero, out of the Boca Chica Country Club and, in the early years, provided English and Spanish versions of basic information for tourists as well as articles on history, business, and general interest topics.[17] According to the masthead, the quarterly magazine was distributed free of charge in the airports of the Caribbean and in the best hotels, restaurants, theaters, and travel agencies across the country. In the first several years circulation was around ten thousand copies but quickly increased to twenty-five thousand before the end of the 1960s. The first few issues, coming in at around fifty pages, included pieces to orient the visitor to the island's particularly storied history, the capital's many historic monuments, and towns outside of Santo Domingo (the first issue even included a map of the country's location in the Caribbean), as well as a list of important telephone numbers and a place for readers to record their "best day in Santo Domingo." However, the magazine also frequently highlighted what still needed to be done to improve the industry. The Camineros were unafraid to criticize the tourism bureau's failure to follow through on particular plans and projects, as they too imagined the nation as a potential tourist mecca.

The third issue of *Bohío Dominicano* (spring 1967) featured a piece called "Alliance for Tourism" ("Alianza para el Turismo") that posed a question: "Can the Dominican Republic become the greatest center of tourist attraction on the American Continent?"[18] The authors, clearly referencing President John F. Kennedy's Alliance for Progress program, bemoaned the "painful reality that has relegated us to oblivion in the world" and proposed an all-encompassing collective for tourism development, to be made up of persons of "moral solvency" who could fund and seek backing for the creation of a robust industry capable of showcasing the "most rich and beautiful corner of the world" that "Columbus most loved."[19] Like their colleagues in the tourism bureau, the editors of *Bohío* pushed to construct a facilitated discovery for potential investors and travelers and to showcase the Dominican Republic's incredible promise as a tourism fantasy.

Creating the "greatest center of tourist attraction" in the hemisphere of course required a significant amount of work, plus loans and legislation to kick the project into gear. In response to these calls, the regime pushed infrastructure expansion through international loans and private investment slowly through the late 1960s. Several pieces of legislation in the late 1960s and early 1970s gave the industry a more substantial base from which to work. The first piece of legislation (Decree 2536), passed in June 1968, formally declared tourism to be of significant national concern, arguing that it was in the nation's best interest to "develop the tourism industry in the Dominican Republic, with emphasis on international tourism," and declaring it obligatory for all entities of government to "coordinate all actions and use of resources concerning or directly affecting the tourism industry, as relevant to the development policies of the tourism industry of the Dominican government."[20] Shortly thereafter, in 1969, the tourism bureau was more formally constituted as a national-level entity (the Dirección Nacional del Turismo, DNT), reporting directly to the president under Law 541, the Organic Law of Tourism. A second piece of legislation, Law 542, created the Corporation for the Development of the Hotel and Tourism Industry (Corporación de Fomento de la Industria Hotelera y Desarrollo del Turismo) to help coordinate tourism promotion and financing with hotel management and general tourism efforts. Finally, in 1971 another piece of legislation, Law 153-71 (the Law of Promotion and Incentivizing of Tourism Development), created the framework for the rapid acceleration of the tourism industry.[21] As a result, the DNT established a series of projects that would benefit greatly from the new legislation; among the projects was the construction of multiple new hotels and resorts.[22] In addition to the tax incentives it created, Law 153 also allowed for significant expropriation of privately held land for development projects, such as Vicini had encouraged in 1967 in his memo to the tourism bureau.[23] Law 153 would continue to serve as the foundational legislation for tourism development, expanded upon by Balaguer's successor, Antonio Guzmán, and through the 1980s and 1990s.

Internal correspondence and annual reports from the National Tourism Bureau in the late 1960s, as well as corresponding press reports, point to a small albeit dedicated core of individuals attempting to demonstrate the nation's great tourism potential, given the new legislation and possible funding sources. They worked to capitalize on the recommendations of the nation's planning bureau by generating private-sector interest in investment opportunities, raising awareness in the general population about the importance of the new industry, building national capacity for visitors, and creating linkages and responding to funding opportunities for tourism through international organizations like the OAS, UNESCO, and World Bank.[24] The numbers of arriving tourists continued to rise steadily, and the regional press slowly began to pick up on the country's potential as a tourism destination. Officials in the public and private sector were largely convinced that the industry was the best economic option for the nation's growth and sought to demonstrate to the private and public sectors that investment in the *industria sin chimeneas* was not only a patriotic and worthwhile cause but one that would pull them away from a primitivistic, dictatorial past and toward a tourism-based future fueled by fantasy.

Re(dis)covering the "Land Columbus Loved Best," 1969-1971

Officials with the National Tourism Bureau used multiple tactics to generate interest in the industry, both from within and outside the country during the last years of the 1960s, and they began to develop marketable concepts of the products on offer to the tourist, including not only the country's natural and human beauty but also its colonial legacy. Bureau directors traveled to different regions of the country to explore the possibilities for development and encourage regional fairs. Folklore and dance exhibitions became regular events, along with a concerted effort to revive extinct or flagging festivals; officials also coordinated "Friendship Weeks" with a number of regional neighbors, including Puerto Rico and Mexico.[25] They created an inventory of the country's best attractions by region and began developing plans for employee professionalization programs in the capital and Puerto Plata. Suitable hotels for travelers were increasing in number, albeit slowly, as were flights into the capital's main airport, soon to be called Las Américas. Outside advertisements, by airlines and major industry players, began to showcase the possibilities of the island's eastern side as well.[26] Much of this work centered on the colonial tropes of rediscovering the "land Columbus loved best."

The capital, Santo Domingo, featured heavily in plans for colonial fantasy-making, as it was the oldest city in the Americas and the site of Columbus's first viable settlement. The vision of Santo Domingo as a city of so many "firsts" in the New World, including the "first beacon of light in the American darkness," was the centerpiece in efforts to brand the Dominican Republic as a place of rediscovery.[27] No doubt heavy-handed and rather an oversell given the conditions of the colonial capital, the showcasing of the colonial center sought to prove incontrovertibly that the island was indeed the "land loved best" by Columbus. Projects to renovate the famous Alcázar de Colón had only just begun, and significant repairs were needed for the cathedral, various colonial churches, and other extant buildings dating to the early 1500s. Still, tourism officials believed these landmarks held significant appeal in themselves but could also entice travelers to the island who might then venture beyond the capital. Foreign consultants, who nearly without fail stressed the centrality of the colonial legacy to tourism, reinforced this strategy regularly.

Support for the bureau's efforts to make the Dominican Republic a showcase for tourism came from local advocates, but the industry gained its most weighty endorsements and directives for future marketing from foreign consultants. As a 1968 UNESCO report had argued, Santo Domingo was the "birthplace of a new culture, the hispanoamericana," as well as a cultural treasure that should be cultivated for global visitors and locals alike.[28] Overlooking the fact that such a narrative ignored the civilization of people (Taíno) that the Spanish trespassers had literally destroyed, the goal was to sell the capital, as well as its surrounding beaches, as the *cuna* (cradle) of American civilization. Spanish consultant Juan Arespacochaga y Felipe was impressed with the historic offerings of the capital city, calling it a "brilliant exception" to the general conditions of the country and contending that each of the many monuments there was its own "spring of tourism attraction."[29] Officials worked hard to incorporate these foreign consultant reports into their campaigns to entice visitors to the "first city." In addition to physical and infrastructural projects, they focused on improving services at the existing capital hotels and expanding the number of offerings for potential visitors to Santo Domingo. Regardless of the city's readiness for the tourism hordes officials desired, Columbian-era monuments and the discourse of discovery proved vital to the burgeoning industry.

Literally tracing the footsteps of Columbus in the capital city was essential to proving the island's authenticity as the birthplace of Hispanic-American culture. In addition to demonstrating that

36. Postcard showing La Ceiba de Colón. Centro León.
Elizabeth Manley collection.

Columbus's heirs lived on and ruled the island in the early sixteenth century from the Alcázar, touristic site promoters sought to demonstrate the very marvel of the explorer's landing, even if the initial crash had occurred far away, on the island's north coast and several years prior to construction of the Alcázar. In one postcard image the "tree of Columbus" (La Ceiba de Colón) was vaguely identified as the site where the seafarer had tied up his ships. The implication was that the three famous ships—the *Niña*, *Pinta*, and *Santa María*—had been tied to the tree, even if those sailing vessels were long gone when the Spanish explorers had made their way to the south coast of the island and founded Santo Domingo in 1496.[30] These locations, as well as all the remaining early sixteenth-century sites, were implicitly connected to Columbus, particularly the first cathedral, in front of which his statue was erected in the late nineteenth century. Formerly the Plaza Mayor, the square and added monument were renamed Parque Colón in 1897. This figurative and literal possession by Columbus—a marker for the island's authenticity as the most historic place in all of the Americas—appealed to government tourism officials as a way not only to set the island apart from other island vacation spots but also to sell the process of rediscovery as much as the place.

The private sector also relied on the imagery of Columbus to sell their products to potential visitors. In the most striking (and rather humorous) employment of the colonizer, the Hertz rental agency implored potential visitors to avoid the "mistakes" made by Columbus. The ad proclaimed "Chris Missed Out on a Lot" and urged visitors to get a car to explore the island.[31] As the copy indicates, not having had such reliable transportation, the island's first "tourist" had not gotten to see all he could have. Other business enterprises, ranging from airlines and travel agents to banks, hotels, restaurants, and various local businesses, similarly engaged the vision of Columbus, Columbus-related locations, or the idea of discovery to sell their products.[32] Iberia Airlines exhorted potential clients to "'Discover' the shortest route to the Old World," while Dominican Air encouraged individuals to "Discover the charms of Santo Domingo"; a local travel agency juxtaposed the Statue of Liberty with the Alcázar and advised that "traveling is the most pleasant way of gaining knowledge."[33] The notion of rediscovery—like Columbus but with modern conveniences—marked many advertisements imploring potential visitors not to miss their opportunity to be the first to uncover the vast possibilities—and paradises—of the Dominican Republic.

Perhaps most notably, the Dominican tourism press, developing slowly through the late 1960s and into the early 1970s, became very reliant on the nation's colonial history to sell the island to the Global North. *Bohío Dominicano* offerings saw the capital and its colonial heritage as huge marketing boons. As the only tourism publication in the early years of industry development, *Bohío* was a major contributor to the efforts to create a tourist paradise in the Dominican Republic for foreign consumption. Despite a relatively low volume of tourists and considerable national skepticism about the viability of the industry, the editors maintained a nearly flawless production schedule, producing the hundred-page-plus magazine four times a year almost without fail.[34] From the beginning, the publication also had the support of some of the industry's biggest players, including the most prominent hotels and restaurants in the capital, the country's air and sea lines, and many local and foreign banks. Like them, they relied on the country's colonial legacy to sell both the magazine and the country; in one early cover illustration the capital's first hospital is shown, with the flowery description arguing that even in ruins it demonstrated that "Santo Domingo, the oldest capital in the American Continent, is the urn holding the relics of a glorious past and the foundations of the somewhat unknown nickname 'Athens of the New World.'"[35]

Those ruins, many in the tourism industry argued, were both crucial to drawing tourists to the island and proof of the nation's blessed place in the tourism game. *Bohío Dominicano* in 1970 praised government officials for their attention to the ruins of the old city.[36] The article "The Ruins Will No Longer Crumble" spoke of the efforts to restore a number of the key sites of the colonial zone. As the city's store of "colonial riches" was one of the propellers of the "tourist ship," inhabitants were already beginning to feel "the onslaught of the first breezes." More "tangible and alive" in Santo Domingo than in other places, the "soul of the past" lived on in the capital in its historic buildings (or ruins), then under renovation by the Office of National Patrimony. It was more than just the age of the buildings, tourism advocates insisted, that made the colonial zone special. Citing President Balaguer, the article reminded readers that the capital was in fact a "vast garden close to the sea, through which angels pass by sounding hosannas on their silver trumpets."[37] Harkening back to Miolán's claim that the deities intended the nation for tourism, advocates for the industry's growth drew on such religious rhetoric to further their claims for even greater investment.

Of course coexisting with the blessed "urn" of colonial relics were the pristine beaches that ringed the island and the sea access touted by Balaguer. While not yet the primary concern for tourism officials, the vast beaches of the Dominican Republic also held considerable draw for would-be tourists, and officials made efforts to build up the facilities in these areas as well. Boca Chica, the beach area closest to the capital, was the most likely target for development efforts, although officials also began to draw up plans for Puerto Plata, a nineteenth-century north coast maritime hub. Yet even in the discussion of beaches the call to explore and discover "like Chris" proved a most attractive marketing tool. It was implicit in the pristine and near deserted beach-scapes of advertisements and marketing materials and more explicit in the invitations to travelers to seek out this new, exciting, and as yet undiscovered paradise. In pushing what would become "heritage tourism," Dominican officials grabbed at what they saw as the country's most distinguishing feature, a feature that was supported by international backers and was the perfect path to guide visitors to this not yet exploited island. The call was out for new tourists to chart a path to the Dominican Republic and for the building of the so-called miracle: the *industria sin chimeneas*.

"What Induces Most Men to Travel to Foreign Lands"

While tourism officials certainly sought to demonstrate how tourists could rediscover the natural resources that made the Dominican Republic ideal for a fantasy escape, climate, landscape, and beaches often played second fiddle to what officials saw as even more crucial: Dominicans themselves. From the very beginning of their efforts, tourism officials understood that gaining the support of the nation's citizens was crucial to the industry's success.[38] As they noted in a 1971 annual report, the Dominican people were naturally hospitable, cordial, and friendly with foreigners, often treating visitors better than they would fellow citizens.[39] More significantly, the nation's female population—the "bellezas naturales del Caribe"—were perhaps the most important inducement to travel. Tourism officials, private enterprise, and the press all supported the belief that beautiful Dominican women could serve as the best ambassadors for the nation's tourism project.

Claiming Dominican women as the ambassadors for tourist discovery was a multilayered project in which beauty pageants featured prominently. The very first year of their development efforts, the National Tourism Bureau revived and celebrated the Dominican Beauty Contest, which had been an annual event. The contestants, representing the various regions of the country, competed for the first three places, named for Dominican products and patrimony (Miss Sugar, Miss Merengue, Miss Coffee), and for the chance to represent the country at international competitions.[40] As *Bohío Dominicano* reported, the event, held in the sumptuous gardens of the Embajador Hotel, was, "without a doubt, another step in the development and promotion of tourism."[41] In the ensuing years, bringing in regional winners, sending Dominican representatives to international competitions, increasing attendance and attention, and expanding their repertoire of pageant-related events became an important part of the tourism bureau's work.[42]

Serving globally as ambassadors of the Dominican brand, contest winners began competing in the regional Miss Tourism (Turismo) of the Caribbean and Central America pageants, as well as at the Miss World and Miss Universe pageants, and they were regularly sent to represent the DNT at various activities and events nationally and globally. After a Dominican won the regional Miss Tourism title in 1970, *Bohío* editors noted their pride but also their optimism for the future of tourism. "With her scepter, crown, and graciousness," editors noted, "Margarita Henríquez inaugurates her reign designed to include her Caribbean and Central American subjects in the great family of international tourism."[43] The full-page image showed the new queen (in a very short skirt) sitting on a throne, and that image was superimposed above a hand-drawn map of the Caribbean and Central America with rays projecting outward—or inward if you will—from Santo Domingo.[44]

Bohío Dominicano proved to be the ideal vehicle in which to showcase the Dominican beauties to both a local and an international audience. The magazine annually, and often more frequently, pictured the Dominican beauty contests, replete with gushing captions for each contestant. In one image a pageant winner seems to stand upon the ocean, holding herself upright with a sunken tree branch, while the copy notes that she offered proof that "there were more beautiful things here [in the Dominican Republic] than on the moon."[45] The message of these images, their boastful captions, and the showcasing of Dominican pageant winners was unmistakably linked to the ideals of discovery embedded in the tourism project, but this overall messaging also signified the role these women were to play as guides and ambassadors to that discovery.[46] As editors noted in 1968 after the regional Miss Tourism contest, held in the Dominican capital, "Latin American countries sent to Santo Domingo what induces most men to travel to foreign lands. *Bohío* is proud to decorate its pages with the beauties in question."[47]

There was hardly an issue of *Bohío* from the 1960s and 1970s that did not feature at least one Dominican beauty queen. Often dressed in evening gowns or even bikinis, the women were pictured in photographs captioned tellingly about their role as tourism ambassadors. The magazine was completely upfront about this tactic, arguing that "year after year, we have adorned *Bohío*'s pages and written about one overwhelming quality of the Dominican woman: her beauty."[48] In the article "A Woman Is Always News" the magazine reported on the 1972 Miss Tourism of the Caribbean and Central America contest, the main feature in the Festival of Merengue/Tourism Week events held at the Embajador Hotel, noting that more than just spectacle, the event was aimed centrally at promoting tourism "through an ambassadress able to strengthen the ties of friendship and increase understanding between Latin American countries."[49] The role of the "ambassadress" in Dominican tourism development was to be both enticing and welcoming for the visitor about to embark on a journey of discovery.

Connecting the country's "natural" beauties and its colonial heritage helped tourism officials sell their Edenic isle. As noted in a caption to a photograph of the Alcázar in the *Bohío* article about the colonial zone's restoration, "this view gives travelers a chance to take their imagination back to the XVI century when Queen María de Toledo and her retinue of maidens . . . used to take their daily walks."[50] Strolling with the maidens—or the beauty pageant winners—proved central to all that the industry was seeking to sell. Private industry sold the nation's unsurpassed natural beauty right alongside tourism officials' promotion efforts. One Holiday Inn advertisement in the late 1960s encouraged visitors to relax "amidst the natural beauty that no other Caribbean country possesses."[51] The goal—to connect the nation's natural beauties with its colonial heritage and possibilities for discovery—was on display whenever possible. As *Bohío* noted, the Dominican Republic was a nation "whose natural beauties and vivid history no other country in America [could] claim to possess," and it yet was "still relatively unknown to international tourism."[52] It was a collectively constructed and totally romanticized image of a past of "firsts," imbued with an objectified female beauty of the present, all in order to create a tourism fantasy for the future.

Over the first half decade of its existence the National Tourism Bureau demonstrated an intense commitment to making tourism a viable component of national economic development, and the bureau was supported in that effort by the private sector, foreign consultants, and the burgeoning beauty pageant business. Despite their limited numbers and resources, individuals and collectives within the industry worked to create linkages with the press and international agencies, mine knowledge from other Caribbean and Latin American tourism projects, and slowly build the bases of a national tourism consciousness.[53] However, it was really at the beginning of the 1970s, particularly with the creation of a funding arm (INFRATUR) for the DNT (through the Banco Central) that national tourism plans and projects began to take on more substance. Underlying the entire project, however, was the powerful discourse of rediscovery and exotic beauty. Joining the colonial and Columbian legacy with the "otherworldly" appeal of the nation's pageant queens solidified the foundations of the tourist endeavor in the Dominican Republic.

Building a Tourism Fantasy, 1972–1978

If the National Tourism Bureau's annual reports from 1967 to 1971 tentatively sang the praises of tourism as a potential player in the national economy, their year-end summaries beginning in 1972 became full-blown gospel on the *industria sin chimeneas* as the miracle cure for Dominican development. Despite their struggles, tourism functionaries insisted that their efforts were highly beneficial to the nation's peace and progress, and they highlighted the ever-increasing numbers of tourists coming to the Dominican Republic.[54] Indeed, the numbers were climbing rapidly. In a distinct shift from 1966, by 1974 the Dominican Republic could boast of 250,000 annual visitors and more than 300,000 if the count included émigré Dominicans returning for a visit.[55] Bureau officials worked aggressively to capitalize on these gains, particularly through suggestions, requests, and recommendations that would be read directly by President Balaguer. Their goal was to make the Dominican Republic a premier tourism destination for middle-class travelers. Creating such a niche in the Caribbean, however, demanded increased financial and legislative attention from the highest levels of the regime. From 1972 through 1978 Dominican tourism officials and their supporters in the tourism press further solidified their emphasis on selling the nation as a discoverable fantasy through the intertwined discourses of colonial legacy and female beauty.

Reaching out to foreign journalists and travel facilitators was crucial in this project of selling the tourism fantasy and its journey of discovery. Inaugurating what travel agents often refer to as "fam trips" (familiarization trips) in 1973, the DNT invited a number of travel agents and writers to come to the island to "sense and feel our development" in the realm of tourism.[56] They argued that such personal linkages were a cheap and effective key to success.[57] For example, they lauded the visit of John Scofield, an editor at *National Geographic*, noting his forthcoming article on the "Discovery of America," expected in a few short months.[58] Although the piece was ultimately called "Christopher Columbus: The Sailor Who Gave Us the New World," the intended effect of the feature article was the same.[59] Readers were reminded of the Dominican Republic's integral place in the explorer's story, and they were encouraged to visit his places of discovery for themselves. Along with similar coverage in other venues, such articles highlighted the island's central nexus in the narrative of "discovery" and the implicit or explicit possibilities for travelers to engage in this same way.

Local publications remained similarly committed to this agenda of beauty and discovery. By the mid-1970s *Bohío Dominicano* had become virtually an arm of the tourism bureau, even if it continued to push the bureau and government to do more. Ten years along, *Bohío* was now distributing approximately twenty-five thousand copies of the quarterly. Pristine beaches, beautiful, bikini-clad women (mostly pageant queens), and historic colonial architecture dominated many of the magazine's advertisements and feature articles. Meanwhile, the editors of the magazine had joined with several other publishers and writers to create the Dominican Association of Tourism Press (Asociación Dominicana de Prensa Turística, ADOMPRETUR). In addition to corralling all the tourism press in the country, *Bohío* continued to promote the fated destiny of the nation and its trajectory en route to becoming a tourism mecca.[60] As President Balaguer argued in 1975, the decision to focus on tourism as a key to national growth was justified "because we have faith in our future as the richest country in the Caribbean area as far as historical and natural beauty are concerned."[61]

Faith in a glorious tourism destiny centered on informing potential tourists of the island's incredible riches while reminding them that they should be "first to discover" this untapped resource for themselves. Many industry promoters, including the Dominican Tourist Information Center in New York, used the "up-and-coming" claim as an inducement to travel and discover the island immediately. Suggesting that "all of its beauties have not yet blossomed," one promotion encouraged visitors to experience the paradise "not yet" discovered by "the Jet Set."[62] Others focused on divine destiny, sometimes in concert with recent renovation efforts. In one short piece a minister's blessing from the 1974 Central American Games contended that "He [God] loves this land far more than Columbus ever could."[63] Throughout the late 1970s advertisements and features highlighted the renovation of the island's resources and offered tempting glimpses into the country's tourism development, all the while stressing the unique and storied history of the Dominican Republic. Such juxtaposition was encapsulated in advertising phrases like "A New Era for an Old Island" or "The New Caribbean Hotel with Old Caribbean Charm."[64] Aware that they still had work to do, promoters encouraged intrepid travelers to take a risk for what they promised would be a high yield.

Bohío served as something of a guide for tourists, as evidenced by the letters regularly published at the end of each issue.[65] However, there was not an official tourism guidebook for the country's intrepid visitors until 1977. An initial attempt at a guide, published in 1972 by a company called Servicios Turísticos, provided some basic information for the visitor.[66] However, for whatever reason, the bimonthly guide did not continue to be produced; in 1975 the Central Bank tried its hand at a guidebook (*Una bella isla del Caribe*), but that also was a one-shot effort. It seems visitors must have survived with *Bohío* and other informational materials provided by hotels and restaurants. Eventually, and somewhat logically, Luis and Rita Caminero took over in 1977 with their *Official Guide to the Dominican Republic*, which continued to be printed annually through the 1980s and 1990s.[67] In their opening message to actual and potential visitors the editors argued that "the friendliness of a land as of yet undiscovered by the mass of tourists will be an unforgettable memory."[68]

This first official guide for tourists in the Dominican Republic sought to highlight the country's current realities and future potential, and it continued the trope of discovery and beauty. It drew on the materials from *Bohío* magazine as well as the several guides that had been published previously. In addition to providing basic background on the country and its history, maps, and lists (of hotels, restaurants, nightclubs, banks, post offices, and churches), the editors sought to extend a warm welcome to visitors. The island was the "best kept secret in the Caribbean," they claimed, because of the affectionate and open nature of its citizens: the "chivalry of our men and the grace of our women [are] traditional."[69] The dual focus on colonial legacy and pristine natural beauty was evident on the cover of the guide, showing both the massive Columbus statue situated next to the national cathedral in colonial Santo Domingo and a typical palm/beach shot. The editors closed the guide by authorizing visitors to "say that this is really an earthly Eden, and that we are willing to share with the rest of the world the beauties that Mother Nature has granted us."[70]

Finalizing the importance of discovering this "earthly Eden" in the Dominican tourism narrative was the formulation of an official slogan in 1975. Despite struggling to find an appropriate tagline for the tourism project in the early years of the regime, officials had by the regime's end finally landed upon an idea that encapsulated many of their efforts to that point. In declaring that "Columbus was right," combined with a silhouette of his statue, Dominican tourism officials cemented their goal of engaging the nation's colonial legacy and building a desire to rediscover the first colonial settlement in the Americas and interact with its "natural" beauty.[71] Taken together, the dual mantras of discovery and beauty indelibly stamped the tourism project as it moved through the final years of the Balaguer *doce años*.[72]

Selling the beauty and Edenic qualities of the Dominican Republic remained at the forefront of development plans, even as much infrastructural growth went on in the background. As the 1970s were ending, the tourism messaging centered even more closely on the display of Dominican women. In discussing the growth of the country's beauty pageants, tourism officials argued for the contests' crucial role as an attraction for tourists. Even if the nation could boast beautiful natural resources, infrastructure, and a colonial legacy, tourism officials argued, "a country without beautiful women would not be well equipped to develop a tourism industry."[73] Tourism promoters across platforms argued that beauty pageants continued to be of special value in tourism development and that women, specifically Dominican women, were "one of the most effective means of promoting tourism."[74] This point was clearly not lost on the editors of *Bohío*, who continued to feature photographs of the country's beauty pageant contestants prominently and throughout the year. The magazine also provided exclusive coverage of both the pageants and the international travels of the pageant winners.

The final three years of the Balaguer regime represented marked advancements in tourism for the Dominican Republic. New partnerships with Club Med, American Airlines, and Sheraton, for example, began to bring both a new clientele and a higher tourism profile. Magazine articles with headlines like "Santo Domingo Combines the Old and the Modern," "Dominican Republic Jumps to First Place in Caribbean Tourism," and "The Oldest City in the Caribbean Looks Brand New" highlighted the industry's advancement as the New World's oldest settlement, now a bastion for tourism. Officials argued that their efforts to make personal connections, their involvement in international organizations and conferences, their fam trips, and, most important, "the peace and prosperity of Balaguer" had made it all possible. Of course considerable loans with ample restrictions, as well as significant private investment, were also major factors. Still, by 1977 the annual number of visitors to the Dominican Republic had reached nearly half a million, representing a more than thirteen-fold increase from the beginning of the Balaguer regime. There was a formal and well-established tourism guide for visitors, and the Santo Domingo airport had received a significant facelift to welcome them. However, what had been most crucial to these advances, even if it remained in the subtext, was the promotion of colonial heritage and rediscovery, as well as the highlighting of Dominican women as ambassadors.

The last full year of Balaguer's *doce años* turned out to be a banner year for tourism, and the hosting of the Miss Universe contest that summer—a major victory for the DNT—was the zenith of the bureau's efforts to use pageant winners to sell tourism. Reporting on its "incalculable impact," officials argued that the global contest was both a demonstration of the work completed so far to advance tourism in the Dominican Republic and an incredible opportunity for further growth. It was estimated that five hundred million people would watch the CBS-televised spectacle, and the host country was given nearly thirty minutes' worth of free advertising time during the broadcast. According to tourism officials, the event, which was executed flawlessly, demonstrated to the world that the country was "a constant carnival of sunshine, verdure, and happiness for life."[75] Reporting on the lavish affair, *Bohío* announced that the global contest, dubbed "our date/rendezvous with the world," unquestionably gave the Dominican Republic "the most valuable international publicity it has ever received."[76] Upon meeting the contestants, President Balaguer proclaimed their importance to the country's growing tourism industry, telling them they would "carry away lasting impressions of Dominican cordiality and the beauty of the country."[77] Ultimately hoping to erase any vestiges of negative impressions from the nation's three-decade dictatorship, tourism promoters welcomed the glowing reviews of the Miss Universe event, which showcased years' worth of the development work, the rightful place of the country in the world of both pageants and tourism, and the future of unmitigated growth in the *industria sin chimeneas*.

By the end of Balaguer's twelve years, tourism had become a key fixture in the development plans of the Dominican Republic, well in advance of the nation's having achieved its status as the most popular tourism destination in the Caribbean. The work undertaken by the National Tourism Bureau, as well as the private sector and the press, had created a solid infrastructural foundation on which the next administration, that of Antonio Guzmán, would continue to build. Most significantly, the growing industry had coalesced around several key tropes to sell the Dominican Republic's virtues to potential visitors from around the globe. Utilizing Columbus and the nation's position as the first Spanish settlement in the Americas, coupled with the unsurpassed beauty of its women, tourism officials generated a tourism fantasy that was available for potential visitors to discover. It was at once "discoverable," exotic, and scripted for exploitation by foreigners, as well as welcoming, guided by beautiful women, and not (yet) overrun by hordes of tourists.

Regardless of differences in methodology and focus, the administration of Guzmán and those that followed the *doce años* would be forced to contend with these entrenched tourism precedents. That is, tourism in the Dominican Republic would remain balanced on key discourses that included a romanticized version of the colonial legacy, an exotic rendering of the nation's female inhabitants, and a travel experience of discovery accessible to and exploitable by all visitors. While most people think of tourism in the Dominican Republic as not truly taking off until the 1980s, when it virtually exploded, the twelve years of the Balaguer regime were crucial to the industry's development, even if there were few actual projects to show for its efforts. Much more significantly, the tourism industry laid the foundations—both logistically and rhetorically—for the future development of tourism for the nation. The tourism script that was sketched out in the early years of the industry's growth in the Dominican Republic represents a first layer in the development of the discoverable island fantasy that was—and in many ways still is—the manner in which the island is sold to the rest of the world.

Notes

1. Quoted in Gregory, *Devil behind the Mirror*, 11.

2. Initially organized in 1934 under Trujillo (Ley No. 4378, del 30 de noviembre de 1934), the Dirección General del Turismo was responsible for tourism affairs and was under the supervision of the Secretariat of Communications and Public Works. The agency bounced around and was renamed multiple times until being reorganized as the Dirección Nacional de Turismo (DNT, National Tourism Bureau) in 1969; I will use the term National Tourism Bureau or DNT from this point forward to avoid confusion. See "Historia," Ministerio de Turismo, accessed August 13, 2019, http://www.mitur.gob.do/index.php/sobre-nosotros/historia.

3. Taylor, *To Hell with Paradise*, 37.
4. Cohen, *Take Me to My Paradise*; Gmelch, *Behind the Smile*; Pattullo, *Last Resorts*; Ward, *Packaged Vacations*.
5. Sheller, *Consuming the Caribbean*; Thompson, *Eye for the Tropics*.
6. Babb, *Tourism Encounter*; Merrill, *Negotiating Paradise*; Schwartz, *Pleasure Island*; Strachan, *Paradise and Plantation*.
7. Strachan, *Paradise and Plantation*; Wilkes, *Whiteness, Weddings, and Tourism*.
8. Trouillot, *Silencing the Past*, 115.
9. Dirección Nacional de Turismo, *Memoria general del año 1970*, 1970, 4, Legajo 17935, AGN.
10. The year before Balaguer's reinvigorated tourism bureau began its efforts, just 27,948 visitors were recorded arriving the Dominican Republic for tourism purposes; in 1967 that number rose to 45,486. Dirección General de Turismo, *Memoria general del año 1967*, 1967, 38, Legajo 17936, AGN.
11. In their 1967 annual report the tourism bureau boasted that they had just "crossed a line away from what might be considered primitivism in order to join this modern phenomenon that is tourism." Dirección General de Turismo, *Memoria general del año 1967*, 1967, 4, Legajo 17936, AGN.
12. José D. Vicini, "Promoción y venta de terrenos dominicanos," memo to Dirección General de Turismo, January 1967, Legajo 116463, Secretaría de Estado de Turismo, AGN.
13. Vicini, "Promoción y venta de terrenos dominicanos."
14. José Andrés Aybar Castellanos to Angel Miolán, November 22, 1967, Secretaría de Estado de Turismo, Legajo 116463, AGN.
15. "Caribbean Vacation Planner," *Caribbean Beachcomber*, March–April 1967, 49–64, ACSC. In contrast, Puerto Rico warranted a full page. The guide was adapted from a British West Indies Airways biannual brochure to include territories "not served by BWIA" and encouraged readers to "take a world tour . . . right here in the Caribbean."
16. Collin Hay, Director, Pan American World Airways, Inc., to Luis Julián Pérez, May 23, 1968, Secretaría de Estado de Turismo, Legajo 116353, AGN.
17. The Camineros had advocated for the industry fiercely since the 1960s; the magazine, now a glossy, full-color monthly, continues to be run by the family, with Rita Cabrer as the editor-in-chief and their son Reynaldo as the executive director.
18. "Alianza para el turismo," *Bohío Dominicano* 3 (Spring 1967): 1, 24, 28, 30, 44, 57–59.
19. "Alianza para el turismo," 1. Most Dominicans were well aware of funds coming into the country through Alliance for Progress programs, although the monies were directed primarily to educational efforts, generalized public works, and rural development. See Rabe, *Most Dangerous Area*, 45.
20. Joaquín Balaguer, Decree 2536, June 1968, Secretaría de Estado de Turismo, Legajo 116353, AGN.

21. Ley 153-71, June 4, 1971, "Ley de Promoción e Incentivo del Desarrollo Turístico," *Gaceta Oficial*, no. 9232 (June 1971), AGN.

22. Portoreal and Morales, "Análisis crítico," 42.

23. Freitag, "Tourism and the Transformation," 231. In addition, as Amalia Cabezas notes, "through World Bank loans and development packages, the productive structure of the country was transformed and its economic strategy redirected toward absorbing foreign investment in tourism. Tax concessions that amounted to more than 10 years of tax exemptions for investment in tourism development were established by Law 153-71." Cabezas, "Tropical Blues," 28. The creation of INFRATUR in 1972, the tourism development arm of the Dominican Banco Central, solidified this economic transformation and shifted focus toward advancing fundable projects.

24. While the influx of funds from international sources like the World Bank and the Inter-American Development Bank dictated where and how infrastructural development would advance, there seemed to be little direction in the way of marketing or advertising.

25. The network of transnational alliances constructed through this process deserves its own study, but it is important to note that Dominican officials reached out to places they believed they should be emulating, including Puerto Rico, Jamaica, and Mexico. Cuba was decidedly not on that list, as the Balaguer regime sought to demonstrate to the United States that there was no chance of it drifting in a leftward direction in the wake of the 1965 occupation.

26. Hay to Pérez, May 23, 1968.

27. "To FLATO," *Bohío Dominicano* 10 (Spring 1969): 1.

28. Juan A. Arespachochaga y Felipe, "Desarrollo Turística, República Dominicana," UNESCO, 1968, 2, BJPD. A mere four years later UNESCO would begin the process of designating World Heritage Sites to spotlight endangered historical spaces around the globe, thus creating a cartography of historical knowledge and discovery, although the colonial city of Santo Domingo was not elevated to that designation until 1990. For more on the impact of the designation, see Ahmad, "Scope and Definitions of Heritage"; and Smith, "Critical Evaluation."

29. Arespachochaga y Felipe, "Desarrollo Turística, República Dominicana," 14.

30. "La Ceiba de Colón," CL, accessed August 13, 2019, https://en.centroleon.org.do/CL/mediateca-publicaciones/3325-coleccion-de-postales.

31. Hertz advertisement, "Chris Missed Out on a Lot," *Bohío Dominicano* 12 (Autumn 1969): 47.

32. Businesses included Iberia, First National City Bank, Banco Popular, Bank of America, Holiday Inn, Fonda de la Atarazana, and legal consultants Medina y Asociados. See issues of *Bohío Dominicano*, 1968–71.

33. These slogans appeared in issues of *Bohío Dominicano*, 1966–70.

34. The earliest issues of *Bohío* give a reasonable sense of what visitors to the island in the late 1960s might have seen and experienced.

35. "Nuestra Portada," *Bohío Dominicano* 5 (Winter 1967): 4.

36. Augusto Ogando, "The Ruins Will No Longer Crumble," *Bohío Dominicano* 15 (Summer 1970): 33–36, 85–87.

37. Ogando, "Ruins Will No Longer Crumble," 87.

38. Such pleas for support continued into the next decades, as evidenced by extremely weathered signs in the small fishing village of Bayahibe in the early 2000s exhorting locals to "wear your life jacket, protect the tourist, be a good example" and "don't litter, smile at the tourist, develop your country." Photographs in possession of the author.

39. Dirección Nacional de Turismo, *Memoria general del año 1971*, 1971, 10, Legajo 17935, AGN.

40. By 1971 four queens had been chosen; see "The Beauty Contest," *Bohío Dominicano* 19 (Summer 1971): 31. The annual event had started in the late 1920s, been revived formally under Trujillo in 1952, and finally was revamped significantly in the late 1960s and early 1970s under the National Tourism Bureau. The Department of Folklore and Beauty Contests (later Concursos y Eventos), a subsection of the bureau, counted on the organizational assistance of a local advertising agency, Ramírez de la Mota, for this first pageant. See "Reinas de Bellezas," *Bohío Dominicano* 6 (Autumn 1967): 4–9. For more on gender and regime pageantry, see Derby, "Dictator's Seduction."

41. "Reinas de Bellezas," *Bohío Dominicano* 6 (Autumn 1967): 4.

42. Through the end of the decade the pageant continued to grow steadily in popularity. The contest pivoted slightly at the end of the 1960s to become part of the bureau's Tourism Week, although it continued to be held at the luxurious Embajador Hotel. In 1968 and 1970 tourism officials played host to the regional Ms. Caribbean and Central American Tourism pageant, which was won in 1970 by a Dominican, Margarita Henríquez. Dirección General de Turismo, *Memoria general del año 1970*, 1970, 37, Legajo 17935, AGN.

43. *Bohío Dominicano* 15 (Summer 1970): 9.

44. A similar image, featuring a bikini-clad pageant winner, had been featured the previous year in an advertisement/editorial. With the headline "Enjoy It," the caption read, "There is a country in the Caribbean area geographically located between 17 degrees 30 minutes north and 20 degrees north, north of the equator and 68 degrees 20 minutes west and 72 degrees west, west of the Greenwich meridian, whose natural beauties and vivid history no other country in America can claim to possess." *Bohío Dominicano* 13 (Autumn 1969): 90.

45. *Bohío Dominicano* 12 (Autumn 1969): 101.

46. They were also not so subtly sexualized. Describing the impact of the contestants in the 1970 Miss Tourism pageant, one *Bohío* editor wrote, "Acting on the public and the jury like a new and heady wine, the candidates' charms looked more accessible under the clinging embrace of the swim suits." *Bohío Dominicano* 15 (Summer 1970): 113. Topless models, if some artistic or medical conceit could be contrived, were also featured in *Bohío* and *¡Ahora!* magazines in the 1960s and 1970s.

47. *Bohío Dominicano* 11 (Summer 1969): 57.
48. "Praise to the Queens," *Bohío Dominicano* 23 (Summer 1973): 1.
49. "A Woman Is Always News," *Bohío Dominicano* 22 (Winter 1972): 2.
50. Ogando, "Ruins Will No Longer Crumble," 86.
51. Holiday Inn advertisement, *Bohío Dominicano* 13 (Winter 1969): 1.
52. *Bohío Dominicano* 13 (Autumn 1969): 90–92.
53. By the end of 1971 tourism officials claimed a tripling of the numbers of available hotels for visiting tourists and noted considerable improvements in existing state offerings. Dirección Nacional de Turismo, *Memoria general del año 1971*, 1971, 55, Legajo 17935, AGN.
54. Officials complained regularly, and sometimes directly to Balaguer, of insufficient budgets, ill-conceived projects, and a less-than-efficient administrative structure. The tourism bureau itself was structurally reorganized several times over the twelve years, evidently in an effort to create greater efficiency.
55. For a number of reasons that seemed to enrage tourism officials, "Dominicanos ausentes" (returning Dominican citizens) were not counted as tourists officially despite the fact that their behavior was nearly identical to that of foreign visitors and they represented a huge component of the department's efforts.
56. Dirección Nacional de Turismo, *Memoria general del año 1974*, 1974, 4, Legajo 17932, AGN.
57. Dirección Nacional de Turismo, *Memoria general del año 1974*, 1974, 4.
58. Dirección Nacional de Turismo, *Memoria general del año 1974*, 1974, 5, Legajo 17932, AGN.
59. John Scofield, "Christopher Columbus: The Sailor Who Gave Us the New World," *National Geographic*, November 1975, 584–625.
60. Octavio Amiama Castro, "A New and Promising Tourist Destiny," *Bohío Dominicano* 40 (Spring 1978): 86–87.
61. Joaquín Balaguer, statement of February 27, 1975, reprinted in *Bohío Dominicano* 29 (Spring 1975): 90.
62. "La República Dominicana: People Like It for What It Is Not," *Bohío Dominicano* 34 (Autumn 1976): 113.
63. Rev. Leonard L. Beard, "Message," *Bohío Dominicano* 33 (Spring 1976): 76–77. In this same issue the editors printed a letter from Rhadamés Trujillo, son of the former dictator, praising the magazine for the promotion of the "industria sin chimeneas" in "nuestra añorada país" (105). It was featured amid similar letters from across the Caribbean and the globe.
64. Gulf & Western Hotels advertisement, *Bohío Dominicano* 35 (Winter 1976): 15; Hotel Hispaniola advertisement, *Bohío Dominicano* 29 (Winter 1977): 20–21.
65. A photographic example of the magazine's use by tourists was captured by photographer Milvio Pérez in the mid-1970s. The picture showed a tourist from New York napping with the magazine resting on her chest. *Bohío Dominicano* 22 (Winter 1972): III.

66. In addition to an overview of the capital and a brief country history, it listed the island's beaches (although not how to get to them), the major hotels, notable restaurants, nightclubs, and colonial zone attractions, as well as lots of other useful information like flight schedules and banking details, postal rates, hospitals, and embassies/consulates. It also contained an article on President Balaguer's significant contributions to tourism development and several additional pieces on Puerto Plata, the 1972 Tourism Week and Merengue Festival, and of course the nation's beauty contests. *Guía turística Santo Domingo* (Santo Domingo: Servicios Turísticos S.A., 1972).

67. Dirección Nacional de Turismo e Información, *Memoria general del año 1975*, 1975, Legajo 17932, AGN.

68. *Official Guide to the Dominican Republic*, 1977 (Santo Domingo), 1.

69. *Official Guide to the Dominican Republic*, 1977, 1, 48.

70. *Official Guide to the Dominican Republic*, 1977, 48.

71. Dirección Nacional de Turismo e Información, *Memoria general del año 1975*, 1975. For the accompanying logo (a copy of the Columbus monument by the National Cathedral), see the official program for the 1976 Merengue Festival, Dirección Nacional de Turismo e Información, "Festival del Merengue," July 20–26, 1976, Legajo 116353, AGN.

72. Although it is unclear when "Columbus was right" ceased to be an "official" slogan, its impact is clear in its continued usage. The website for the town of Jarabacoa in the central mountain region of the country boasts, "When Christopher Columbus, the Dominican Republic's first tourist, stepped ashore on this land over 500 years ago, he marveled at its beauty. Writing to King Ferdinand and Queen Isabella, he called it 'The most beautiful land the eyes of a man have ever seen.' To this day, there is no record of any visitor ever having contradicted him. . . . Once you've visited the Dominican Republic, you may also agree that Columbus was right. It is certain that no other country could hope to match the profuse splendor of its intensely green valleys punctuated by scores of sparkling rivers, nor equal the awesome majesty of its towering mountains, nor rival the primitive beauty of its cactus filled deserts, nor compete with the silvery allure of its innumerable beaches." "About Jarabacoa," Jarabacoa tourism website, accessed August 13, 2019, http://www.mi-vista.com/jarabacoa.htm.

73. Dirección Nacional de Turismo, *Memoria general del año 1972*, 1972, 27, Legajo 17935, AGN.

74. Dirección Nacional de Turismo, *Memoria general del año 1972*, 1972, 27.

75. Quoted on the second leaf of Dirección Nacional de Turismo e Información, *Memoria general del año*, 1977, Legajo 17930, AGN. An ad called the event "nuestra cita con el mundo." *Bohío Dominicano* 37 (Summer 1977): 1.

76. David Ahlers, "Miss Universe 1977," *Bohío Dominicano* 39 (Autumn 1977): 1. The magazine, in a separate article written by editor Luis Caminero, noted the significance of the win by the contestant from Trinidad and Tobago. "Eli-

gen una morena Miss Universo/Election of a Colored Beauty Miss Universe," *Bohío Dominicano* 39 (Autumn 1977): 51–56.

77. Ahlers, "Miss Universe 1977," 3.

Bibliography

Archival Sources

Archivo de Ciencias Sociales y del Caribe, UPR-RP, San Juan, Puerto Rico. Various tourism publications (ACSC).

Archivo General de la Nación, Santo Domingo, Dominican Republic. Files from the National Tourism Bureau and National Development Commission and volumes of the *Gaceta Oficial* (AGN).

Biblioteca Juan Pablo Duarte, Banco Central, Santo Domingo, Dominican Republic (BJPD).

Biblioteca Nacional Pedro Henríquez Ureña, Santo Domingo, Dominican Republic. Repository for *Bohío Dominicano*, *¡Ahora!*, and the *Official Guide to the Dominican Republic* (BNPHU).

Centro León, Santiago, Dominican Republic. Postcard Collection (CL).

Published Works

Ahmad, Yahaya. "The Scope and Definitions of Heritage: From Tangible to Intangible." *International Journal of Heritage Studies* 12, no. 3 (2006): 292–300.

Babb, Florence E. *The Tourism Encounter: Fashioning Latin American Nations and Histories*. Stanford CA: Stanford University Press, 2011.

Cabezas, A. L. "Tropical Blues: Tourism and Social Exclusion in the Dominican Republic." *Latin American Perspectives* 35, no. 3 (2008): 21–36.

Cohen, Colleen Ballerino. *Take Me to My Paradise: Tourism and Nationalism in the British Virgin Islands*. New Brunswick NJ: Rutgers University Press, 2010.

Derby, Lauren. "The Dictator's Seduction: Gender and State Spectacle during the Trujillo Regime." *Callaloo* 23, no. 3 (2000): 1112–46.

Freitag, Tilman G. "Tourism and the Transformation of a Dominican Coastal Community." *Urban Anthropology* 25, no. 3 (1996): 225–58.

Gmelch, George. *Behind the Smile: The Working Lives of Caribbean Tourism*. Bloomington: Indiana University Press, 2003.

Gregory, Steven. *The Devil behind the Mirror: Globalization and Politics in the Dominican Republic*. New ed., with a new preface. Berkeley: University of California Press, 2014.

Merrill, Dennis. *Negotiating Paradise: U.S. Tourism and Empire in Twentieth-Century Latin America*. Chapel Hill: University of North Carolina Press, 2009.

Pattullo, Polly. *Last Resorts: The Cost of Tourism in the Caribbean*. New York: Latin America Bureau, Monthly Review Press, 2005.

Portoreal, Fátima, and Marco Morales. "Análisis crítico de la legislación y las políticas turísticas en República Dominicana." In *Turismo placebo: Nueva colonización turística del Mediterráneo a Mesoamérica y El Caribe; Lógicas espaciales del capital turístico*, edited by Maciá Blázquez and Ernest Cañada, 29–52. Managua, Nicaragua: Editorial Enlace, 2011.

Rabe, Stephen G. *The Most Dangerous Area in the World: John F. Kennedy Confronts Communist Revolution in Latin America*. Chapel Hill: University of North Carolina Press, 1999.

Schwartz, Rosalie. *Pleasure Island: Tourism and Temptation in Cuba*. Lincoln: University of Nebraska Press, 1999.

Sheller, Mimi. *Consuming the Caribbean: From Arawaks to Zombies*. New York: Routledge, 2003.

Smith, Melanie. "A Critical Evaluation of the Global Accolade: The Significance of World Heritage Site Status for Maritime Greenwich." *International Journal of Heritage Studies* 8, no. 2 (2002): 137–51.

Strachan, Ian G. *Paradise and Plantation: Tourism and Culture in the Anglophone Caribbean*. Charlottesville: University of Virginia Press, 2002.

Taylor, Frank Ford. *To Hell with Paradise: A History of the Jamaican Tourist Industry*. Pittsburgh PA: University of Pittsburgh Press, 1993.

Thompson, Krista A. *An Eye for the Tropics: Tourism, Photography, and Framing the Caribbean Picturesque*. Durham NC: Duke University Press, 2007.

Trouillot, Michel-Rolph. *Silencing the Past: Power and the Production of History*. Boston: Beacon Press, 1995.

Ward, Evan R. *Packaged Vacations: Tourism Development in the Spanish Caribbean*. Gainesville: University Press of Florida, 2008.

Wilkes, Karen. *Whiteness, Weddings, and Tourism in the Caribbean: Paradise for Sale*. New York: Palgrave Macmillan, 2016.

FOUR
Postmodern Ironies and Dark Tourism

11

Mina El Edén and Dark Tourism in Zacatecas, Mexico

ROCIO GOMEZ

In 1960, when Mina El Edén ceased operating as a silver mine, the city of Zacatecas regarded the closure as the sign of a passing era. The silver that once buoyed the Spanish Crown was now mined in more rural veins and owned by foreign interests. Despite the economic setback, the region committed to remaking itself as a tourism center soon after Mina El Edén's closure, marketing itself instead as a home of *charros*, or Mexican cowboys, and stunning landscapes. But can a region built on silver turn away from its industrial past, especially when the remnants of extraction rest in plain sight? While such a change is unlikely, a city can reorient this industrial past to its benefit while at the same time controlling the historical narrative surrounding it. Tourism presented an opportunity to generate profit using the skeletal remains of the mine; at the same time, the mine's use as a tourist site provided a narrative that raised the profile of the attraction and the region by association. Moreover, the destination relied on the unique quality of Mina El Edén: a retired sixteenth-century mine only blocks from the historic downtown district. While bustling streets now surrounded the site of centuries of silver ore extraction, the story of the mine presented on tours reimagined the industry as a boon, not as an enterprise detrimental to the environment or dangerous to its workers. Quite the contrary—the tourism experience at Mina

El Edén reiterated mining as the city's heritage without referring to the horrific details of workers' illnesses, accidents, and deaths.

As local mining investment slowed in the mid-twentieth century, municipal officials in the city of Zacatecas scrambled for new economic activity. Tourism at Mina El Edén was one way to attract visitors to the area. With its history as a mining center, the city of Zacatecas reinvented itself as a domestic tourist attraction for Mexicans interested in the city's colonial architecture and its mining history.[1] Just as the floors of Mina El Edén were cleared of debris in preparation for visitors, the constructed tourist narrative sanitized the mining history of the region, ignoring the hazards to workers and the environmental effects while emphasizing the mineral wealth of the region. Coming on the heels of increased foreign ownership of mineral rights in Mexico, the tour emerged as a museum for the ecological remnants of capitalist extraction. On January 1, 1975, the Mina El Edén tourism experience opened to the public, as the state's governor, General Fernando Pámanes Escobedo, promoted the site as a step forward in the region's economy.

This chapter examines the visitor experience at Mina El Edén in Zacatecas to show how tourism reinterpreted the region's relationship to its industrial heritage. I argue that the inclusive definition of "dark tourism" permits a reading of how the growing urban center reconciled its extractive past with its more profitable future as a tourism destination, especially for domestic tourists already aware of the city's importance in the colonial era (1546–1821). A relatively new field in tourism studies, dark tourism permits a reevaluation of how some cities sought to recover from the neoliberal shift away from manufacturing and toward foreign investment and consolidation. Two points of discussion underscore how Mina El Edén emerged as a marketable attraction for the region and how dark tourism affects the historical retelling of extraction. While mining still occurs in cities such as Zacatecas and Guanajuato, both saw growing numbers of visitors in the late twentieth century because of their mining past and not in spite of it. With this shift to tourism, tourism officials and docents alike have embraced the opportunity to shape the narrative and present a sanitized tale to visiting members of the general public.

The Mina El Edén Experience

Visitors to dark tourism sites often have a curiosity about the macabre. Whether the sites are literal boneyards or the skeletal remains of industrial sites, they represent death to a certain degree. In the case of Mina El Edén, tourists take in "death by capitalism" as they visit a retired mine, an industrial remnant in a city famous for its silver production. The city of Zacatecas and its historic silver mine are just one example of how shifting a historical narrative allowed municipal officials to offer their crumbling industrial works as somewhat macabre tourist destinations. Anthropologists and sociologists have analyzed this practice of resurrecting collapsed industry as sites of tourism in the attempt to reorient regional economies.[2] Smokestacks, abandoned power plants, and ghostly steel mills all signify an industrial and manufacturing past that began collapsing across North America in the 1960s. The remains of these industrial parks drew visitors for various reasons, ranging from curiosity about the silent machines to fascination with the science behind manufacturing. During guided tours, visitors admired the giant mechanisms that once powered the golden ages of manufacturing or they gawked at the enormity of steel mills. As sites of curiosity, mines underscored the importance of the industry to the history of the country and emphasized the foundational role mining played in colonial cities like Zacatecas, Guanajuato, or Taxco.

However, the tragedies of workplace accidents and environmental pollution that resulted from mining make the tourism around Mina El Edén an example of "dark tourism." Coined in 1996 by Malcolm Foley and J. John Lennon, the term "dark tourism" refers to the growing interest surrounding sites of tragedy and calamity. Early on these authors acknowledged that the broad definition allowed scholars from various disciplines to analyze and understand the cultural and historical narratives of these areas.[3] While these destinations offered histories that drew visitors, they prove unsettling in their mass appeal. However, Philip Stone defines the

37. C. B. Waite, *Historic Downtown Zacatecas*, ca. 1900. Wikimedia Commons.

term in the context of action: "the act of travel to sites associated with death, suffering, and the seemingly macabre."[4] The physical act of travel to sites associated with death, suffering, or hard labor underscored the motivation that drove tourists' desire to visit these places.[5] Regardless of the reasons surrounding the visits, tourism purveyors cited the growing number of curiosity seekers when federal or municipal officials questioned the validity of promoting tourism at sites that had witnessed pain and tragedy. Chernobyl has become the most famous example of dark tourism. In April 1986 workers at that nuclear power plant in Soviet Ukraine faced a partial reactor meltdown that generated a green mushroom cloud over the nearby city of Pripyat. The Soviet government delayed evacuation of the city, and thousands of families suffered exposure to radioactive materials, which has affected subsequent generations. While tourists and the abundantly curious wandered onto the site on occasion, a gripping 2019 television series reignited interest and drove tourist hordes to the exclusion zone, despite persistently high levels of radiation.

Since opening in 1975, the mine tour in Zacatecas has drawn visitors with the promise of a glimpse into the cavernous remains of the mining legacy in the region. Raymundo Montes, a local artist and business operator, pushed the city to develop the mine as a tourist attraction. While the initial goal may have been to preserve the patrimony of Zacatecas, visitors to the mine brought tourist dollars to a region facing the challenges of a shifting economy.[6] In many cases, *zacatecanos* had long used mines as hideouts, storage facilities, or water-gathering sites, and they had established informal and unregulated entrances into extraction sites around the urban center. When mining moved to the outskirts, the mine in the center of the city became an object of curiosity and a potentially profitable attraction that recalled the mining heritage of the region. Because it was a site that appealed to both foreign and domestic tourists, tourism officials aimed to educate the public on the mineral wealth of the city and its mining heritage.[7]

Even so, tour guides today sanitize the narrative of the mine tour, rendering it largely ahistorical. The whitewashing begins

38. Tourists at silver mine entrance. Photographer unknown. Wikimedia Commons.

before tourists even enter the mine. Upon purchasing a ticket, visitors receive a hairnet and a hard hat, which has been dutifully sprayed with Febreze. Next they board a small train. The mechanism resembles children's trains common in amusement parks, with a miniature engine pulling cars painted like mine carts. Adding to the experience, visitors then pull a Plexiglas bubble over their cart for protection against falling rocks as they are pulled roughly six hundred meters into the mountain. After a seemingly interminable journey through rock, the visitors arrive at an open area several degrees cooler and noticeably drier than the outside air. They then proceed to the geology exhibit, which displays fossils, metals, stones, and gems found near and far. Gazing at the samples secured behind protective glass, visitors marvel at the cleanliness of silver or the brilliant shine of processed copper. A timeline displays the discovery of silver in Zacatecas, the colonial exploitation of the metal, and the process of refining the metal.

Once the guests have seen the geology exhibit, a docent begins a guided tour at the mouth of a lit hallway. Standing on the polished stone floors (as can be seen in images on the mine's website), he explains the history of mining to a captivated audience. The narrative begins with the founding of the city and discusses Mina El Edén during its peak in the nineteenth century.[8] Using props placed along the way, he points to plastic models of indigenous miners hauling sacks of ore strapped to their forehead and pulling the rocks out of the mine with the sacks on their back. Wryly, he notes that the miners proved to be masters of physics, since they performed this task while teetering on wooden and rope ladders dozens if not hundreds of feet in the air. Visitors then hear sound effects of bats when they pass a darkened corner, giving the docent an opportunity to reassure them that the sounds only belong to the cave's other inhabitants. Farther along, tourists come upon an altar, which hosts El Santo Niño de Atocha, the preferred religious icon for miners in the region.

While exploring the cavernous area, tourists receive two contrasting narratives on mining: one via a river visible in the depths of the mine and the other via an out-of-place oddity. Near the end of the tour the docent leads visitors over grates that offer a stomach-churning view of the depths. Strategically placed lights shine on a raging river seventy meters below, which glows a toxic green. Meanwhile, the docent notes the pollution of the river and the undrinkable quality of its water, adding to the horror and the vertigo. He adds that this contamination came from other towns surrounding the state capital and not from the city. In contrast to the river narrative intended to evoke revulsion, a strange sight prompts tourists to strain for a view of it upon arriving on the train: a nightclub. Next to the geology exhibit, the nightclub (appropriately called La Mina) houses a bar and a DJ station. During its hours of operation, a laser show flashes pickaxes, hammers, and images of a mountain on the mine's walls, all while the latest electronic dance hit plays. Giant speakers hang down from the ceiling like would-be stalactites while modern chairs line the dance floor and corridors as if they were boulders emerging from the mine itself.

Analyzing Eden

Mina El Edén offered a degree of entertainment to the tourists drawn to the region by featuring a discotheque and organized surroundings in the remains of an industrial operation.[9] Despite these entertaining diversions, the mine does not escape the label of dark tourism because of the human stories buried in the depths of the mine. With death and suffering an inherent part of the definition of dark tourism, the history of workers' accidents and debilitating occupational diseases challenge the entertaining narrative presented by the tour guide. The workers' stories and the pollution in the river belie the gleaming environment and modern conveniences in the tourist venue that the mine has become.[10] The ahistorical depiction of the mine on this tour ignores the historical evidence of workers' deaths and suffering, as well as the environmental evidence of excavation. Tourists experience a simulacrum of mining that ignores the plight of miners and the large-scale environmental degradation. Despite the hard hat and train ride, visitors have a make-believe experience that dismisses long-standing regional ties to the industry. Specifically, the mine tour captures a brief, singular moment that largely ignores the long accumulating history of extraction. In addition, the overall tone of the tour is lighthearted, in direct contrast to the serious endeavor that is the act of extraction. In effect, the tour guide successfully colonizes the minds of tourists with a specific idea of mining in Zacatecas. The tour sanitizes the impacts of centuries' worth of mining activity in the region; the guides gloss over the lingering toxic legacies of mining and how they affect the health of the general population to this day. Even now the agricultural fields in the region contain significant traces of lead, cadmium, and mercury, according to environmental scientists.[11] Aquifers under the city and throughout the region suffer from chemical and mineral contamination, a dire situation in a state that is increasingly parched. The glorious vision offered by the Mina El Edén experience presents an ahistorical artifact and ignores its current iteration in the Canadian-owned open-pit mines operating just a few miles away.

Dark tourism largely ignores the historical presence of human activity and instead emphasizes the technology of the industrialized past. Sites that attract visitors include abandoned factories, empty mines, and obsolete steel mills, which invite tourists to marvel at the enormous machines that once ran regional economies. However, the stories of the workers who labored at those sites and operated the giant machinery have been largely ignored, as have accounts of their injuries and the occupational hazards of work in that bygone era. Instead of relating the history of people working at the site, the tourist venue often replaces that narrative with one that is more pleasant and palatable to visitors. Dark tourism has preserved the suffering of workers only in the instruments of their torture—coke smelters, drills, assembly lines—while demonstrating and in some ways mourning the death of industry in the region. The collapse of mining and the manufacturing industry shut down not only the physical workplace but also an entire region's economy, throwing it into the boneyard of capitalism and neoliberalism. Workers faced an unfolding tragedy as their livelihoods and hometowns emptied, while their identity, closely tied with the industrial heritage of the area, suddenly lost its footing. How does the loss of industry intersect with dark tourism? A finality serves as the central focus—death identified with a specific site and also with the overall death of industry. The former recalled human mortality, while the latter illustrated a slow decline and ultimate obsolescence for its machines and its workers under the weight of competition from abroad, where stingy corporate bosses could pay lower wages, introduce greater mechanization, and exploit more accessible raw materials.

In many cases the conversion of former industrial sites to tourist attractions succeeded, but mining tourism in particular carried the label of dark tourism, which further complicated not only the presentation of the sites to the public but also the preservation process. Mines offered insight into places of hard physical labor. Unlike steel mills or open-pit mines, which have mechanized operations, mining in Mexico historically relied on blood labor, or the physical activity of humans, in order to extract the ore from the mountains. In the colonial period the workers relied on "hands, hammers, and fingernails," as the common saying goes, in order to loosen the rocks from the mine walls. In Zacatecas workers did not have mechanized equipment until the late nineteenth century, when steam-powered drills finally arrived. Even then, workers faced an additional encumbrance with the clouds of dust that filled the underground shafts and tunnels, where slippery footholds and nearby drops compounded the danger. The carved walls of tourist mines are the result of hard physical labor that could claim men's lives at any time.

Docents presented a narrative that included some details of the labor and presented the chiseled walls as evidence of the arduous task, along with folktales spawned by miners. In many cases tour guides faced the challenge of putting these facts into context while also maintaining tourist interest, which required balancing discussion of the industry, the cultural history of mining, and the heavy cost in human lives. In this case the loss of lives in mining was glossed over in favor of humor and entertainment. Mining, in short, is dangerous. Notoriously unsafe conditions characterized their occupation; miners endured not only taxing labor but hazardous conditions such as falls, cave-ins, occupational diseases, and accidents. Consequently, they created their own mythologies and superstitions surrounding their work, presenting their labor as verging on the supernatural. Trinidad García (1831–1906), a mine owner and politician from Zacatecas, wrote a book recalling the legends of demons and devils in the rocks. In one particular tale a miner lost in the darkness of the mine encountered an eerie voice in the depths. "Don't be scared. I am here with you," it whispered. When he struck a match to see who spoke the words, he peered into the hollow eyes of a skull, a remnant of a former miner.[12] In another local legend a seventeenth-century miner remained trapped in the rock of Mina El Edén. According to the folktale, Roque was a miner considered hefty and rude by his companions. His colleagues tolerated him, however, because, though bombastic, he was a prodigious producer of ore. On one occasion the supervisor asked Roque to start digging in a dangerous part of the mine. While initially hesitant, he agreed and moved to his new post. After digging a short time, he discovered rocks of gold. He was overjoyed, as he always craved being rich and longed to leave the mine behind. Containing his joy, he hid his precious rocks and went to retrieve his bag in order to smuggle them out. When he returned to the loot, the gold rocks were gone. Angry and distressed, he loudly blasphemed and blamed God for turning against him, a poor miner. Soon the walls started to shake and cave in on Roque, trapping him in the rock. On the tour the docent recounts the story of Roque and points to a spot on the wall across a small ravine. The rock forms three-quarters of a startled man's face and a mining helmet. Workers also recorded their suffering and their comrades' deaths in song, adding to the cultural history of the industry. Macario González and Francisco González sang the melancholy lyrics of the *canción* "El Minero" in 1956: "At six, I present myself at the mine / I enter the halls of death."[13] Meanwhile, modern miners in Guanajuato offered prayers to Santo Cristo de los Mineros, a patron saint to the miners of La Valenciana mine.[14] Thanks to prayers and *corridos* composed and repeated about mining, the dangers associated with the occupation recorded themselves on the collective memory of Mexican miners.

Does Mina El Edén qualify as dark tourism? Did tourists visit the mine to hear these macabre tales? Or is it a dark tourism site disguised as a sanitized mine experience? Foley and Lennon offer a broad enough definition of dark tourism to include a variety of interpretations and tourist expectations. Dark tourism at its core emphasizes the historical importance of the site while forcing present-day audiences to face their own mortality. Despite a sanitizing narrative and a question of whether tourists knew the history of extraction, Mina El Edén illustrated dark tourism through its cavernous ruins and anxiety-inducing ambiance. In addition, visitors marveled at how humans, using only muscle power, carved the inside of the mine in an era without steam-powered engines. Both entombed and suspended above dark depths, tourists confronted their own insignificance within the remnants of an industrial site and the enormity of a centuries-old mine.

Eden in Context

With the mine experience, tour guides and city officials wanted to draw a crowd to a relic of the mining past while shaping the narrative of extraction as a glorious endeavor rather than one that was ecologically devastating. In part, tourists absorbed a tale of mineral wealth that enriched the region, all with little context in the form of the larger story of ore extraction in the region. Specifically, they received a history that upheld the wealth and not the horrific circumstances in which workers extracted the valuable metals. Tour guides glossed over the ecological effects of mining, and visitors encountered clean environs in their mine tour. Despite a scrubbing of the complicated history from the presentation, the mine itself illustrated the historical impacts of mining and its material remnants, albeit in an entertaining, ahistorical yarn.

Indigenous settlers in Zacatecas were the first to begin mining in the region, and their activities led to the colonial extractive industry that sustained the Spanish regime with silver. Before the Spanish arrived in 1546, indigenous miners had employed an ingenious technique to expose silver in the hills. Building a fire close to the walls of an overhang or natural cave, early miners heated the surface of the walls before throwing water on them, exposing silver in the cracks that formed in the sudden temperature change.[15] When the Spanish arrived in what is now Zacatecas, an indigenous man gave Spaniard Juan de Tolosa a rock with streaks of silver in it—a gift that ushered in European settlement and recognition by the Crown two years later. Intertwined in this relationship between the inception of the mining industry and the founding of the city was an identity central to the city's historical narrative. The identity of the city and of the mine continued to be largely inseparable because of the spatial relationship of city and mine. With three major silver veins below the city plazas, there existed no distinction between city and industrial center. The veins wound down from the north-northwest, through the city center, and left the valley in a east-southeasterly direction. Various mines began operating along these veins, with entry points cut into a subterranean network that saw extraction sites stack atop each other, chambers next to each other, while a beehive of activity developed as miners followed the veins. The economic importance of the region, as well as the voraciousness of extraction industries, led to a cluster of mines near the city center. Centuries later Mina El Edén drew tourists as a top attraction. El Edén was even close to another prime attraction, the central cathedral, thus doubling the tourist draw to the region.

Docents for the mine tours told a tale of workers toiling in the mine depths, but the narrative glossed over the dangers of everyday work in the mines over the centuries. In the colonial period workers utilized what they could to dig raw ore from the walls. Unlike the forced labor (*mita*) of indigenous people in South American mines, some miners in Zacatecas were being paid wages by the end of the seventeenth century.[16] Mine owners and heads of the viceroyalty of New Spain agreed to pay wages in order to lure skilled workers to the city. Regardless, the city relied on miners to provide the "blood labor" needed in silver mining, despite providing few tools and shoddy engineering. Little changed with regard to extraction methods after independence. Miners continued taking chances in speculative mining as well, pockmarking the surrounding region with extraction areas even as foreign interests began encroaching. As the Porfirio Díaz dictatorship (1876–1910) increasingly invited foreign companies to buy into the mines, the number of miners in the city of Zacatecas grew during the 1880s as silver boomed. During this period miners saw any existing precautions in safety erode, as fierce competition and newer technology pushed extraction activity even deeper into the earth. In addition, new mines opened on the city's outskirts, making the journey to and from populated villages to remote hillside mines dangerous. Even more so, when injuries occurred in these far-flung mines, victims from accidents or falls had to endure a journey of several hours on foot or by carriage, which doomed many in need of immediate medical attention.

Despite the miners' importance to the extraction process, mine owners risked their workers' lives daily in shafts supported only by wood, rope, and luck. Injuries and accidents were commonplace. Workers suffered injuries in falls from extreme heights as they maneuvered the subterranean network of shafts and tunnels on ladders, ropes, and slippery floors. In mine levels close to aquifers, the presence of water or high temperatures in the underground environment, as well as hard labor and sweat, created moisture in the air that collected on walls and made floors slick. The lack of ventilation later contributed to lung diseases, especially in those charged with breaking new ore. Foreign investors in the nineteenth century imported new extraction technology from abroad and thus added another layer of danger: mechanical trauma. Inexperienced workers miscalculated the effects of new machines and extraction technology in the late nineteenth century, leading to a rash of accidents. Some miners used machines incorrectly. Others received poor training in the new technology. Nonetheless, accidents and tragedies occurred every day with the introduction of new technology, yet such details never entered into the mine tour narrative.

While it found success again as tourism site, Mina El Edén had achieved success in its prior use by consistently producing silver in the early twentieth century—and reliably high numbers of worker injuries. On the eve of the Mexican Revolution, the production statistics of Mina El Edén revealed it to be a profitable yet exploitive extraction site. From January to May 1910, Mina El Edén (registered as a property of the Compañía Minera y Beneficiadora El Edén, S.A.) listed its total size as 14.40 hectares (roughly 35 acres).[17] The company employed 100 men and 10 boys for yearly salaries ranging from 3,600 to 5,200 pesos, with a daily take-home pay of 0.34 to 5.00 pesos. In total, these 110 men and boys hauled 2.4 million kilograms (2,646 U.S. tons) in this same time span, with the help of one 80-horsepower steam engine. From this amount, they refined 484,394 kilos of gold, silver, and lead for a value of 18,700 pesos. Along with producing at such a high level, workers suffered a high rate of accidents, as reported in statistics from the same time period. While the statistics do not describe the precise nature of the accidents, 2 miners died in the mines while 31 suffered injuries during the same time period. By comparison, the other successful mine in the city, Mina El Bote, saw 71 injuries, with all workers recovering.[18] The mining overseer recorded these injuries in the company ledger alongside the production statistics, with human labor listed as either productive or sidelined due to injury. The reduction of humans to statistics exemplified the mining heritage of the region as well as the relationship between the city of Zacatecas and the industry. This statistical categorization underscored the invisibility of workers despite their impact on the land and the economy.

While docents for the mine tour did not mention workers' struggles, the two cases described below represent a fraction of those detailed in the archival records. While these accidents did not occur in Mina El Edén, they are representative of the injuries common in mining during the peak years of production at Mina El Edén. In addition, municipal officials considered workers' injuries so severe and violent that they launched criminal investigations to determine if there had been foul play, which led the investigation paperwork to be placed in the criminal records section of the Archivo Histórico del Estado de Zacatecas. In the first case Félix Hernández worked in Mina El Diamante in February 1891, nearly eight miles from the center of Zacatecas, according to the legal memorandum attached to the incident.[19] Hernández, in excavating the mine, placed dynamite in holes bored through the rock in order to expose new ore veins.[20] He quickly exited the mine in order to let the fuse reach the charges. Growing impatient for the blast, he went to relight the fuse, thinking the flame had gone out. As he approached the fuse, the explosion went off, and he received the brunt of the blast. Adding to the trauma, the blast wave drove a drill bit into his chest. Coworkers quickly notified the victim's cousin, Pascual Hernández, a laborer in the same mine, who ran to the scene and dragged the young man out into the open air. Félix died soon after. Miners often experienced a variety of physical traumas due to the increased use of steam engines, dynamite, and machinery. Distance often compounded the injuries, as most happened far from hospitals and medical care. Coworkers of Félix Hernández made a valiant effort to take him to the hospital, but he did not arrive in Zacatecas from Vetagrande until six hours after the explosion.

Another frequent type of accident besides the tragedies with dynamite and technology was the almost daily frequency of cave-ins. Berardo Carreón and Francisco Hernández both experienced trauma from a cave-in on October 31, 1891, in a mine located outside Sauceda, a small town ten miles from the state capital. Carreón bore the brunt of the falling rock and died of his injuries, as the autopsy report details. The rocks crushed his midsection, broke ribs on his left side, and smashed his left leg. While the injuries did not kill him instantly, he surely died from the blood loss, per the examiner's description. Hernández narrowly escaped with a broken leg and shattered pelvis. As with Félix Hernández and now Carreón, municipal authorities called for an investigation to determine who, if anyone, should be held responsible for the deaths of these workers, asking victims' colleagues to provide witness testimony.[21]

With mining taking place in the city and with the regional economy dependent on it, the collapse of silver markets on three separate occasions—1825, 1893, 1910—reiterated the reliance of the city on the industry. This reliance illustrated why city officials agreed to make the mine experience central to the reinvention of Zacatecas, a city that had faced the need to reimagine itself repeatedly when the silver market crashed. In 1825, after Mexico had gained independence from Spain, mining collapsed because Spaniards had owned many of the mines. Those who left Mexico to return home had little choice but to abandon excavation sites, especially if they felt wary living in the new republic. Without investment, mine operators lacked the substantial capital needed to maintain a worksite. In 1893 the bottom fell out once again due to the Sherman Silver Purchase Act of 1890, passed by the U.S. Congress. Faced with tumbling prices on the global silver market, Zacatecas investors soon found extraction operations too financially risky to continue, leading to the closure of mines in the region along with the shuttering of businesses in the state capital. In the twentieth century new investors drove a renaissance in mining until the adoption of large-scale earth-removal technology combined with an economic boom drove many miners to larger cities for work. After 1910 the military phase of the Mexican Revolution swept through the region. Soldiers purposely flooded mines and others destroyed equipment after the Battle of Zacatecas in June

1914. While business fled the city, miners and foreign investors found footholds in silver mining in the hinterlands of the state in an attempt to hide from roving armies. With mining becoming increasingly costly, the slow demise of silver mining as well as the "Mexican Miracle" in the 1950s made urban areas more attractive for opportunities and better salaries, drawing workers away. Furthermore, miners saw local jobs disappear as multinational companies began accumulating mining sites throughout the state and incorporated even more earth-removal technology, further displacing workers. Thus, the decline of mining began with the closing of mining sites within the city in the early twentieth century and the dissolution of domestic mining operations in the face of competition from foreign corporations. As a result, the city began to pivot toward a more tourist-driven economy with the marketing of its famous *charreada* (rodeo) and the Morisma de Bracho, a reenactment of battling Christians and Moors.

Conclusion

Tourists frequent dark tourism sites out of curiosity or morbid attraction, and their frequent targets are abandoned factories and mines. However, these sites often present information in an ahistorical manner that neglects the dangers and suffering as well as the factors that led to the site's or the industry's collapse. The Mina El Edén tourist experience in Zacatecas ignores the historical hazards associated with mining and keeps the narrative as neutral as possible, glossing over the ecological hazards of extraction still present today in the region. Dark tourism scholars have underscored the role of death, hard labor, and suffering in drawing tourists to specific destinations. Yet, because mining tourism replaces a collapsed industrial economy, docents present a history of extraction that has been scrubbed of historical truths about the inherent dangers or the work and the way mining has left a toxic legacy in the region. Thus, in order to draw visitors, tour operators offer only a glimpse of the mine and its past, one that is scrubbed and sterile, omitting any mention of the suffering, toxic legacies, and death begat by the industry.

Notes

1. For colonial histories of Zacatecas, see Bakewell, *Silver Mining and Society*; and Velasco Murillo, *Urban Indians*. For a modern environmental history of the city, see Gomez, *Silver Veins, Dusty Lungs*.

2. For a comparative discussion of industrial heritage, mining tourism, and dark tourism, see van Veldhoven, "Post-Industrial Coal-Mining Landscapes"; Michels, *Permanent Weekend*; Ávila-García, Luna Sánchez, and Furio, "Environmentalism of the Rich"; Rothman, "Stumbling toward the Millennium"; LeCain, *Mass Destruction*, 188–200; Steel, "Mining and Tourism"; Ferry, "Memory as Wealth"; and Ferry, "Inalienable Commodities."

3. Foley and Lennon, "Editorial: Heart of Darkness," 195–97. See also Lennon and Foley, *Dark Tourism*.

4. Stone, "Dark Tourism Spectrum," 160.

5. Stone, "Dark Tourism Spectrum."

6. Ownership of the derelict mine became an issue when interest in promoting the site for tourism began to develop. See Roberto Ponce, "Zacatecas, Patrimonio de la Humanidad," *Proceso*, January 29, 1994. He argues that slow economic development had rendered the city full of "poverty and lethargy," a situation that somewhat ironically preserves intact the historic structures of a city, saving them for new life as tourist attractions.

7. See the Mina El Edén website, http://www.minaeleden.com.mx/english/.

8. Some tour guides are women. However, the author of this chapter has taken the tour twice, and on both occasions the same man led the tour.

9. In 2013 local media reported that the previous state governor, Ricardo Monreal Ávila, had sold the mine to a private interest. José Aguirre Campos, the head of a local company, reiterated that the mine had never belonged to the city and that it had always been managed by private companies. Manuel Frausto, "Monreal no vendió mina El Edén: Dueño responde a Pepe Bonilla," *Zacatecas Online*, November 11, 2013, https://zacatecasonline.com.mx/noticias/local/34771-monreal-no-vendio.

10. Sharpley, "Shedding Light on Dark Tourism."

11. Pearson, "Preliminary Findings"; Iskander, Vega-Carrillo, and Manzanares Acuna, "Determination of Mercury," 45–48; Zetina Rodríguez, "La controversia ambiental," 167–71.

12. García, *Los mineros mexicanos*, 74–75.

13. González and González, "El Minero."

14. Ferry, *Not Ours Alone*, 111.

15. Brown, *History of Mining*, 6–8.

16. Bakewell notes that some miners were enslaved. Bakewell, *Silver Mining and Society*, 122.

17. Mine Ledger, Exp. 2, Caja 17 (1911), Estadísticas, Jefatura Política, AMdeZac.

18. Mine Ledger, Exp. 2, Caja 17 (1911).

19. Lic. F. Henriquez, memorandum, February 13, 1891, Exp. 537, Caja 21, Serie Criminal, Judicial, AHEZ.

20. Lic. F. Henriquez, memorandum, February 27, 1891, Exp. 537, Caja 21, Serie Criminal, Judicial, AHEZ.

21. Crescensio Ríos, testimony before Lic. Jesús Zamora, October 31, 1891, Exp. 575, Caja 22, Serie Criminal, Judicial, AHEZ.

Bibliography

Archival Sources

Archivo Histórico del Estado de Zacatecas, Zacatecas, Mexico (AHEZ).
Archivo Municipal de Zacatecas, Zacatecas, Mexico (AMdeZac).

Published Sources

Ávila-García, Patricia, Eduardo Luna Sánchez, and Victoria J. Furio. "The Environmentalism of the Rich and the Privatization of Nature: High-End Tourism on the Mexican Coast." *Latin American Perspectives* 39, no. 6 (2012): 51–67.

Bakewell, Peter J. *Silver Mining and Society in Colonial Mexico: Zacatecas, 1546–1700.* Cambridge: Cambridge University Press, 1971.

Brown, Kendall W. *A History of Mining in Latin America: From the Colonial Era to the Present.* Albuquerque: University of New Mexico Press, 2012.

Ferry, Elizabeth Emma. "Inalienable Commodities: The Production and Circulation of Silver and Patrimony in a Mexican Mining Cooperative." *Cultural Anthropology* 17, no. 3 (2002): 331–58.

———. "Memory as Wealth, History as Commerce: A Changing Economic Landscape in Mexico." *Ethos* 34, no. 2 (2006): 297–324.

———. *Not Ours Alone: Patrimony, Value, and Collectivity in Contemporary Mexico.* New York: Columbia University Press, 2005.

Foley, Malcolm, and J. John Lennon. "Editorial: Heart of Darkness." *International Journal of Heritage Studies* 2, no. 4 (1996): 195–97.

García, Trinidad. *Los mineros mexicanos: Colección de artículos sobre tradiciones y narraciones mineras....* México DF: Oficina de la Secretaría de Fomento, 1895.

Gomez, Rocio. *Silver Veins, Dusty Lungs: Mining, Water, and Public Health in Zacatecas, 1835–1946.* Lincoln: University of Nebraska Press, 2020.

González, Macario, and Francisco González. "El Minero." *Mexican Corridos.* Washington DC: Smithsonian Folkways Recordings, 1956.

Iskander, Felib Y., Hector René Vega-Carrillo, and Eduardo Manzanares Acuna. "Determination of Mercury and Other Elements in La Zacatecana Dam Sediment in Mexico." *Science of the Total Environment* 148, no. 1 (May 1994): 45–48.

LeCain, Timothy J. *Mass Destruction: The Men and Giant Mines That Wired America and Scarred the Planet.* New Brunswick NJ: Rutgers University Press, 2009.

Lennon, J. John, and Malcolm Foley. *Dark Tourism.* London: Thomson Learning, 2004.

Michels, John. *Permanent Weekend: Nature, Leisure, and Rural Gentrification.* Montréal QC: McGill-Queen's University Press, 2017.

Pearson, Ron. "Preliminary Findings in Assessment of Soils and Crops in the Zacatecas Area, Mexico." In *North American Regional Action Plan for Mercury: Close-Out Report.* Montréal: Commission for Environmental Cooperation, May 2013.

Rothman, Hal K. "Stumbling toward the Millennium: Tourism, the Postindustrial World, and the Transformation of the American West." *California History* 77, no. 3 (1998): 140–55.

Sharpley, Richard. "Shedding Light on Dark Tourism: An Introduction." In *The Darker Side of Travel: The Theory and Practice of Dark Tourism,* edited by Richard Sharpley and Phillip Stone. Bristol, England: Channel View, 2009.

Steel, Griet. "Mining and Tourism: Urban Transformations in the Intermediate Cities of Cajamarca and Cusco, Peru." *Latin American Perspectives* 40, no. 2 (2013): 237–49.

Stone, Philip. "A Dark Tourism Spectrum: Towards a Typology of Death and Macabre-Related Tourism Sites, Attractions, and Exhibitions." *Tourism: An Interdisciplinary Journal* 54, no. 2 (2006): 145–60.

van Veldhoven, Felix. "Post-Industrial Coal-Mining Landscapes and the Evolution of Mining Memory." In *Landscape Biographies: Geographical, Historical, and Archaeological Perspectives on the Production and Transmission of Landscapes,* edited by Jan Kolen, Hans Renes, and Rita Hermans, 327–44. Amsterdam: Amsterdam University Press, 2015.

Velasco Murillo, Dana. *Urban Indians in a Silver City: Zacatecas, Mexico, 1546–1810.* Stanford CA: Stanford University Press, 2016.

Zetina Rodríguez, María de la Carmen. "La controversia ambiental en torno a la presa La Zacatecana, Guadalupe, Zacatecas." *Desacatos* 51 (May–August 2016): 160–79.

12

Netflix *Narcos* and Narco-Tours

Film Tourism Meets Dark Tourism in Medellín, Colombia

FÉLIX MANUEL BURGOS

The first time I felt the violence of Pablo Escobar was in May 1990, when a loud explosion shook the windows of our house in Cali, Colombia. The next day we learned from the news that it was a car bomb, put in a drugstore called Drogas La Rebaja. In Cali it was an open secret that Drogas La Rebaja belonged to the Rodríguez family, leaders of the Cali cartel. This attack would be the declaration of war by the Medellín cartel against their rivals in Cali. The violence of drug trafficking had spread like an uncontrollable fire throughout Colombia, and the name Pablo Escobar would become synonymous with fear and impotence.

Many years later, in my classes at New York University, the name Pablo Escobar reemerged. From semester to semester I would ask my students about famous people from Colombia. Pablo Escobar was the first and in many cases the only answer. When mentioning his name, my students would knowingly look at each other, laugh, and murmur. Through these interactions I came to realize that there was a fascination with Escobar and that they did not associate him with the kind of fear he once struck in me. They referred to him as a kind of drug dealer hero who challenged the law while living a life of luxury, extravagance, and violence. When inquiring about the origin of this particular representation, I received answers that were nearly unanimous: the Netflix series *Narcos*.

This chapter stems from this curious observation. I'm interested in exploring the effect of *Narcos* on the representation of Pablo Escobar among international audiences and how this has affected the tourism industry in Medellín, Colombia. To this end I examine the boom in Narco-Tours, which coincides with the broadcasting of the popular Netflix series. I propose that this is a case in which dark tourism and film tourism converge and that the narratives of fiction end up complementing and in some cases replacing history. My discussion concludes with a reflection on how Colombians have reacted to this phenomenon in terms of national identity in the global sphere.

Narcos

The *Narcos* series was released by Netflix in August 2015. The show features a sophisticated production, with locations in Colombia, where the life of Pablo Escobar, played by the Brazilian actor Wagner Maoura, is re-created. A Drug Enforcement Administration (DEA) agent's voice-over narration presents the historical context against original footage of Pablo Escobar and Colombia in the 1980s and 1990s. This factor is key to crafting a documentary feel to *Narcos*, as selected historic images are strategically included in the production of the drama to create a sense of authenticity. Further enhancing the show's appeal to North American audiences, the use of a DEA agent's voice-over provides viewers a familiar "war on drugs" tie-in to the Colombian setting.

The characterization of Pablo Escobar in *Narcos* includes a personal angle. As the series unfolds, he is portrayed as a relatively quiet and unassuming man in the context of his family and personal relationships while at the same time he is working to build the most powerful drug cartel in history. This is a captivating but hardly original narrative in film or television. The same sort of formula can be seen in, for example, *The Godfather* (1972), *The Sopranos* (1999–2007), and *Breaking Bad* (2008–13). Like these other successful film and television dramas, *Narcos* similarly establishes a dynamic tension between the vulnerabilities of Escobar's private life and the violent power and prestige of his professional operations.

39. Pablo Escobar mug shot. Wikimedia Commons.

The series, propelled by Netflix's role in global media, has rekindled international interest in Pablo Escobar and in the recent history of Colombia. This is reflected in searches for the name Pablo Escobar on Google, where its highest search frequencies coincide with the premieres of the first and second seasons of *Narcos*.[1]

The international popularity of Pablo Escobar cannot be attributed exclusively to the launch of the Netflix series. Before the show's pilot premiered, in mid-2015, dozens of books and documentaries reported on his life and criminal empire. Still, the release of *Narcos* has introduced Escobar to a new generation of viewers, the majority born after his death in 1993.

In Colombia the phenomenon is not entirely different. Escobar's rise and fall are largely associated with a history of kidnappings, car bombings, assassinations, and violence. In response, there have been several fictional portrayals of *narcotráfico* that have subsequently created new layers of representation and interpretation. In Spanish these dramatizations are known as *sicaresca*, a term derived from the noun *sicario* (hit man) and the term "picaresque," a literary genre harking back to sixteenth-century Spain. The sicaresca depicts a dark world of unscrupulous characters who operate in a society corrupted by organized crime. In addition to a number of novels and short stories that sketch the sicaresca, there are a variety of film and television productions in this genre, such as the Colombian *telenovela* titled *El patrón del mal* (2012).

The Netflix series *Narcos* introduces the sicaresca to audiences across the globe while portraying Pablo Escobar as a heroic figure. Ambitious and daring, Escobar managed to overcome social barriers to become an international drug dealer. While transgressing mainstream social hierarchies, Escobar is at the same time portrayed as a family man, with a particularly loving relationship with his mother. This quality gives his character an emotional appeal that is then complicated by his relationships with other women, such as his wife and a number of attractive, affluent women with whom he has affairs.

In addition to the heroic characterization of Pablo Escobar, the appeal of *Narcos* is also driven by an action-film-style dramatization of violence. In the series, violence is presented strictly for its presumed entertainment value and not from a social documentary or historical perspective. As a result, the devastating impacts of Escobar's operations and the drug wars more generally on Colombian society are not considered. The intentional omission of the ethical and social costs of drug-related violence is a determination that will—as we shall see—subsequently inform the realization of Narco-Tours in Colombia.

In summary, there are three elements that contribute to the romantic portrayal of Pablo Escobar in the Netflix series. First, his personal life, although difficult to verify from a historical point of view, is dramatized in an appealing manner. Next, the violence of the Medellín cartel, and of the drug wars more generally, is framed as action-adventure entertainment, which then allows the audience to view tragic and deadly acts from a comfortable distance. Finally, the voice-over narration of the series by a U.S.-based DEA agent inscribes the narco saga within the larger, hegemonic perspective of the war on drugs.[2] This perspective allows North Americans to view the series with a sense of legitimacy by which they interpret Escobar's life from the lawful or "right side" of the story.

The Narco-Tours

Tourism in Colombia has increased considerably in recent years. In 2016 the number of tourists visiting the country totaled 1.9 million, twice as many as in 2005, and by 2018 the Ministry of Commerce, Industry, and Tourism had reported more than 4.2 million visits by nonresident foreigners.[3] This sustained growth of tourism is directly related to a decrease in violence, which is largely a result of the 2016 peace agreement between the Colombian government and the former guerrillas of Fuerzas Armadas Revolucionarias de Colombia (FARC). As violence waned, tourism emerged as an important economic area of growth in Colombia, especially in urban centers such as Bogotá, Cartagena, and Medellín. Known to locals as the "city of eternal spring," Medellín has seen a steady increase in tourism and anticipated as many as 2.5 million national and international annual visitors in 2019.[4]

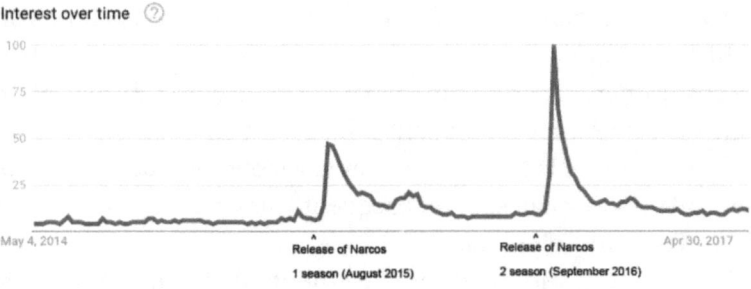

40. Narcotourism graph, 2014–17. Félix Manuel Burgos.

In Medellín, tourism generates considerable employment and is generally viewed in a positive way by local residents. However, some tourist activities have generated controversy and in some cases spurred vocal rejection by concerned citizens. The most criticized form of tourism is *narcotourism*, which consists of travelers visiting areas of the city where they can consume drugs and even visiting plantations or clandestine drug laboratories. Narcotourism is also associated with sexual exploitation, whereby prostitution of adults and minors is packaged alongside the consumption of illicit substances.[5]

On the other hand, there are travel agencies that offer tourist plans known as Narco-Tours. Unlike narcotourism, this activity is legal and does not involve, at least explicitly, the use of drugs or prostitution. These are guided tours of places associated with the Medellín cartel and specifically with the life and death of Pablo Escobar.

Tours of this nature have existed since before 2014.[6] However, since the beginning of mid-2015, the year Netflix released the series *Narcos*, there has been a significant increase in the number of agencies offering such experiences. Although it is impossible to establish a causal relationship, it is reasonable to presume a connection between the popularity of the series and the increased demand for

41. Narco-Tour packages. Félix Manuel Burgos.

such tours. A factor that offers evidence of this connection is the use of images from the series in the design of promotional materials created by various companies selling Narco-Tours.[7]

The proliferation of Narco-Tours alongside the appearance of the series can be understood within the theoretical framework of film tourism, which is "a specific pattern of tourism that drives visitors to see screened places during or after the production of a featured film or a television production."[8] Most of the literature about this topic is based on fictional productions, such as *The Lord of the Rings*, which spurred tourism in New Zealand.[9] However, the case of the Narco-Tours is more complex, given that the story portrayed in the series is based on real events and people. This tension between reality and fictionalization can then be considered a convergence of film tourism and dark tourism—two complementary forces that attract visitors and, at the same time, two theoretical dimensions that allow observers to understand the phenomena.

The concept of dark tourism has been used to explain the fascination people develop for places associated with death.[10] However, new studies have lent dark tourism a broader dimension by examining the intentional creation of such spaces. As A. V. Seaton

42. Colombia narco hoodies. Félix Manuel Burgos collection.

proposes, "Dark tourism/thanatourism compromises encounters through travel with the engineered and orchestrated remembrance of mortality and fatality."[11] This notion is crucial if we are to understand the emergence of Narco-Tours in Medellín. Unlike historical places that commemorate death in terms of remembrance or homage to the victims (e.g., sites associated with the Holocaust or 9/11), the Colombian Narco-Tours are highly orchestrated visits to places identified with Pablo Escobar. In this sense the death and destruction of the drug lord's victims are not the main focus of interest; the place and the generation of violence itself are the appeal.

Violence as a focal point of tourism has been documented from different perspectives.[12] One motivation for tourists is a desire to experience the sensation of danger while avoiding actual harm.[13] In the case of Narco-Tours, the very act of visiting Medellín can in and of itself evoke a feeling of fear and danger, given that the city has in recent years been associated with crime, violence, and terrorism. Understanding this, the travel agencies offering Narco-Tours do so in a way that both capitalizes on Medellín's reputation and maximizes visitors' sense of safety and comfort.

As of 2019, some Medellín travel agencies were offering a wide variety of Narco-Tours, with prices ranging from $30 to $400 (U.S. dollars). With the most expensive plans, excursionists are provided

43. Tourists and guide in Medellín. Félix Manuel Burgos.

the opportunity to interact with some of the actual people involved with the Medellín cartel, such as Pablo's brother, Roberto Escobar, alias El Osito. While most traditional agencies don't offer these types of tours—presumably because they wish to avoid giving their businesses a bad reputation among locals—there are small new agencies dedicated exclusively to Narco-Tours.

Ethnography of a Narco-Tour

I took a Narco-Tour in Medellín in 2018. The venture cost sixty dollars, and upon receiving my payment, the agency indicated that I should meet the tour guide in the El Poblado neighborhood, which is an upper-class area of Medellín. At the meeting point there was a white minivan waiting. Our group consisted of five foreigners and myself, the lone Colombian. The tour was run by the driver and a guide, both of whom gave us a warm welcome as we climbed into the van's cool air.

Jonathan, our guide, was a typical *paisa*.[14] Charismatic and talkative, with decent facility in English, he supplemented his presentation with words in Spanish when necessary. After a brief presentation, one of his first questions to the group was, "Who likes *Narcos*?" Two members of the tour enthusiastically raised their hands, while the others politely nodded. Our guide explained that we were going to enjoy an "authentic experience," because he had not only grown up in the city but had even gotten to know several of the old members of the Medellín cartel.

This claim of authenticity was central to a particular strategy of what tourism studies has identified as essential in developing an allegedly "authentic" experience: object-related connections (e.g., original place, physical/historical context) and participant-related connections (e.g., bodily feelings, emotional ties, associations, etc.).[15]

In relating what was advanced as an authentic version of Colombian narco history, our guide Jonathan was keen to point out that we were visiting the actual places where Pablo Escobar's actions took place, not merely those presented in the show. In this assertion there was a double approach to the Netflix series. Initially, *Narcos* was used as a motivator for the travelers to engage with the tour, but then the show was dismissed as a largely fictional construction in contrast to the "authentic" experience we were undertaking in Medellín. In further promoting his "true" account, Jonathan developed a commentary on the drug war in Medellín. This was a discourse articulated in the first-person voice, rife with personal anecdotes that incorporated Jonathan as an actor with emotional ties to the places we were visiting.

Our 2018 tour began at the Mónaco Building, where Pablo Escobar once lived with his family before going into hiding. This destination is no longer available because national and local authorities ordered the building demolished in a public campaign to reform the image of the city.[16] The demolition of the Mónaco is an example of what Kenneth Foote calls *obliteration*, an effort to erase the places associated with an uncomfortable past in order to generate new local and national identities.[17]

The second part of the tour was a stop at La Catedral, the infamous prison where Pablo Escobar and many of the Medellín cartel members served time. La Catedral is also the place where Escobar later staged his famous escape, dramatically re-created in *Narcos*.

Just outside the prison one encounters a highly visible sign with the phrase, "Those who do not know their history, are condemned to repeat it" ("Aquellos que no conocen su historia están condenados a repetirla"). Curiously, this same expression of Spanish philosopher and essayist George Santayana is featured in the introduction of the 2012 telenovela *El patrón del mal*. Although the origins of the sign are somewhat unclear, there is little doubt that the rendering of this iconic quote just outside the famous La Catedral prison is a good example of how mass media can reshape certain places in order to make them more appealing to visitors. In this case deployment of the Santayana quote specifically appealed to Colombians familiar with the popular *El patrón del mal* series.

The third stop was the most emblematic example of dark tourism and the climatic part of the tour: the Jardines Montesacro cemetery in Itagüí, just south of Medellín, where Pablo Escobar is buried. This tour stop is one of the most popular and controversial. As one enters the cemetery, the solemnity of the space is interrupted by the arrival of groups of tourists, often noisy. Their presence generates a tension between visitors wanting to catch a glimpse of the drug lord's burial site and those who are there for more legitimate personal, family, and ceremonial reasons.

El Patrón, as Pablo Escobar was known, lies under a modest tombstone surrounded by images, flowers, and other offerings. In this part of the tour our guide commented on the different versions of his death—from the official story involving an elite military squad known as the Bloque de Búsqueda to theories suggesting that Escobar committed suicide before being captured. Once again our guide invoked the *Narcos* series by explaining that the location where the scene of Escobar's death was shot in the series was just two houses down from the actual site.

At this point of the visit the common activity of taking pictures with cameras and cell phones during the tour became even more frequent. This behavior is suggestive of the intimate relationship between photography and dark tourism.[18] In this, one can imagine that the very act of taking a photo in such a place could be read intertextually as a way in which the photographer is superficially

44. Medellín neighborhood. Félix Manuel Burgos.

able to manifest a connection with a historical or recognized site. Taking photos is not simply a way to create memories of a trip. Thanks to social media, it is also a manner in which "individuals [can] express their identities, and create and maintain social relations online."[19] Being at the grave of Pablo Escobar and posting pictures with hashtags such as #narcos and #medellin allow visitors to present themselves in a variety of ways: as "true fans" of the television show or perhaps even more as brave, intrepid travelers, given Medellín's international reputation as a dangerous place. Amusingly, one enthusiastic member of our tour group, an American male in his early twenties, even commented to his friend that "the best selfie here would be having a line of cocaine on [Escobar's] grave."

The last stop of the tour is the neighborhood La Comuna 13, one of the peripheral areas of Medellín, where various scenes from the *Narcos* series were filmed. Here again, the field of tourism studies helps us to appreciate the constructed power of these urban landscapes by highlighting various mental impressions held by tourists prior to arrival.[20] In this context traveler expectation is part of a larger, three-stage sequence (hypothesis-input-check) proposed by

Clare A. Gunn in his classic 1972 work *Vacationscape*.[21] Initially, the idea of a destination starts with a hypothesis or imagined appearance of the place. For potential visitors to Comuna 13 who are motivated by the show, *Narcos* is a pivotal motivator given that various episodes took place in different neighborhoods in the foothills of Medellín, including Comuna 13. Narco-Tour stops in Comuna 13 then became what Gunn refers to as "input": making physical contact with a place. Once arrived at Comuna 13, fans of *Narcos* were given a detailed commentary by our guide, who referenced specific scenes from the show while pointing to various houses, corners, and other landmarks depicted in the series. This activity further represented Gunn's sequence in realizing a successful "check" in comparing expectation with the input experience.

Another important element of this final stop was the narrative produced around La Comuna 13. According to our guide, this sector of the city was one heavily afflicted by the drug war involving the Medellín cartel. At the same time, however, it was an area where Pablo Escobar built strong popular support by donating money, houses, soccer fields, and other social goods to residents.

Still, this Narco-Tour presentation of the neighborhood site was somewhat misleading. Tracing the history of Escobar's largesse, one learns that it was another neighborhood—La Comuna 9 rather than La Comuna 13—that benefited most from his patronage. It is Comuna 9, not Comuna 13, where there is a neighborhood named after the kingpin. And, as it turns out, Comuna 13 was the site where political violence (i.e., between guerrillas and paramilitaries) rather than cartel violence occurred during the conflict.[22] Our guide, following the institutional discourse, presented La Comuna 13 as a symbol of the transformation that Medellín has experienced since the conflict; it is a city that has gone from being one of the most dangerous places in Colombia to becoming a center of social development and a tourist attraction. Guide Jonathan's closing remarks on the tour were clearly crafted as a metaphor for the tour and the city itself, highlighting as they did a significant change from a dark past to a new resilience and the renewed promise of the Medellín of today.

Conclusion

Narcos has had a significant impact on international perceptions of Pablo Escobar, the city of Medellín, and the nation of Colombia. For many, this show is their main source of information about the infamous Colombian drug dealer of recent Colombian history. Given the characteristics of the production, which combines original footage with staged scenes, this show presents a biographic narrative in which Escobar is portrayed as a heroic international bandit who defies agents of both the Colombian and U.S. governments. This David-versus-Goliath confrontation, richly peppered with sex, violence, and action/adventure, has created an alluring antihero with a contemporary fan base extending across much of the globe.[23]

This fascination with Escobar has also had an effect on the rapidly growing tourism industry in Medellín. The city has more than sixteen agencies that specialize in Narco-Tours largely related to the life and death of the notorious drug lord. In this the realms of film and dark tourism give rise to two important considerations.

First, these tours reflect the new ways in which history is reproduced and consumed. The show presents a version of the history of the Medellín cartel immersed in the sphere of entertainment. This representation leads Colombian tour guides to adapt their narrative to the expectations of foreign travelers and to offer a view that largely satisfies what is presented by the show. Local historical narratives are thereby revised and in some cases replaced by the fictions of global mass media to satisfy consumer demand. In so doing, one can clearly observe ways in which globalization and postmodernism distort national memory.

Second, it is important to consider the reaction and critical perspective Colombians have generated in response to both the *Narcos* series and Narco-Tours. Generally, it can be observed that Colombians have rejected the *Narcos* series. There are any number of examples where discontent with the Netflix series has been made explicit, such as the Facebook group Colombia NO Es Pablo. In this group, which has more than thirty-six thousand members, nearly all aspects of the series are criticized, including its being advertised abroad and the sale of merchandise associated with Pablo Escobar, which is wholeheartedly denounced. Group members argue that there are several reasons to denounce *Narcos*, especially the glorification of Escobar. Further annoyance with *Narcos* was summarized by Juanes, the internationally known musician from Medellín, who shared in an interview just how upset he was by the image of Medellín and Colombia portrayed to the world by *Narcos*.[24]

There is no doubt that drug trafficking is linked to the history of Colombia and has had enormous social, economic, and cultural effects.[25] Yet, despite this reality, Colombians do not want this image to be propagated abroad. The problem here is not so much the historical past but the for-profit romanticization of Colombia's drug cartels for global entertainment. Before the Netflix *Narcos* series, the Colombian-made *El patrón del mal* had indeed proven to be a success among Colombians. What gives rise to the widespread rejection of *Narcos*, however, is a general feeling among Colombians that their "dirty laundry" is now not simply "being washed at home" but in the global marketplace.[26] In contrast to Netflix's *Narcos*, the telenovela *El patrón del mal* was mainly a national product aimed at a Colombian audience—an intimate representation of national affairs in which dirty clothes were appropriately being washed at home.

Criticism of Narco-Tours has similar explanations, although it leads to a more complex problem: repeated travel to places that Colombian authorities and a large portion of the national public would like to leave behind. In recent years, especially with the peace agreement between FARC and the government, there has been an institutional effort to present Colombia, and especially Medellín, as a society that has managed to overcome its problems of the past in favor of a brighter future. Initiatives such as these conflict with the practice of the Narco-Tours and therefore lead to official discourses of sanction and criticism.[27]

Colombians' problem with these types of tours is ultimately related to a problem of collective identity. As tourism industry development and growth are often closely associated with heritage sites and discourses that more positively assert national identity, history, culture, and social meaning, a perceived lack of such history and culture may allow for the exploitation of less savory episodes in the national past. With still so little distance from Colombia's tragic history to provide needed national perspective, the visitation of foreign tourists to "dark sites" of Colombia's recent past continues to discomfit the nation and its people.

Notes

1. Search of "Pablo Escobar" in the United States from December 2014 to December 2016, Google Trends, accessed June 5, 2019 https://trends.google.com/trends/explore?date=2014-12-31%202016-12-31&geo=US&q=Pablo%20escobar.

2. Zavala, *Los cárteles no existen*, 42.

3. MINCIT, "Estadísticas de turismo en Colombia," Ministerio de Comercio, Industria y Turismo (Colombia), 2019, http://www.mincit.gov.co/minturismo/inicio.

4. "Medellín espera 2.5 millones de turistas en 2019," *El Colombiano*, June 4, 2018, https://www.elcolombiano.com/tendencias/turismo-medellin-nuevas-apuestas-YN8807045.

5. UNODC, "Estudio exploratorio descriptivo."

6. Giraldo Velásquez, Van Broeck, and Posada, "El pasado polémico de los años ochenta."

7. "Pablo Escobar Tour—Tours Medellín—Tours—Discover Medellin and Enjoy All of Our Tours," *Medellín City Tours* (blog), accessed May 23, 2019, http://www.medellincitytours.com/pablo-escobar-tour/.

8. Roesch, *Experiences of Film Location Tourists*, 6.

9. Buchmann and Frost, "Wizards Everywhere?"

10. Seaton, "Guided by the Dark."

11. Seaton, "Encountering Engineered and Orchestrated Remembrance," 11.

12. Andrews, *Tourism and Violence*.

13. Lozanki, "Desire for Danger, Aversion to Harm."

14. This is the term used by Colombians to refer to people from the department of Antioquia and its capital city, Medellín.

15. Knudsen and Waade, *Re-Investing Authenticity*, 10.

16. "Pablo Escobar: Qué significa el derribo del edificio Mónaco, uno de los mayores símbolos del poder del narco colombiano," BBC News Mundo, February 22, 2019, https://www.bbc.com/mundo/noticias-america-latina-47286796.

17. Foote, *Shadowed Ground*, 24.

18. Lennon, "Dark Tourism Visualization."

19. Munar and Gyimóthy, "Critical Digital Tourism Studies," 255.

20. Lai and Li, "Tourism Destination Image."

21. Gunn, *Vacationscape*, 29.

22. Angarita Canas et al., *Dinámicas de guerra y construcción de paz*.

23. There is a French Instagram account that serves as an example of the many international sites that offer merchandise associated with Pablo Escobar: "Narcos Medellin (@narcosmedellin93) Instagram Photos and Videos," accessed May 24, 2019, https://www.instagram.com/narcosmedellin93/.

24. "El Hormiguero: Juanes carga contra la serie 'Narcos' por la imagen que da de Colombia," *La Vanguardia*, April 24, 2019, https://www.lavanguardia.com/television/20190424/461851624104/el-hormiguero-antena-3-juanes-la-plata-narcos-netflix-medellin-colombia.html.

25. Rincón, "Narco.estética y narco.cultura en Narco.lombia."

26. This observation comes from the Colombian saying "la ropa sucia se lava en clasa" (dirty laundry is washed at home), which implies that private matters should be kept secret.

27. Van Broeck, "'Pablo Escobar Tourism'—Unwanted Tourism," 291–318.

Bibliography

Andrews, Hazel. *Tourism and Violence: New Directions in Tourism Analysis*. Farnham, England: Ashgate, 2014.

Angarita Canas, Pablo Emilio, et al. *Dinámicas de guerra y construcción de paz: Estudio interdisciplinario del conflicto en La Comuna 13 de Medellín*. Medellín: Sello Editorial de la Universidad de Medellín, 2016.

Buchmann, Anne, and Warnick Frost. "Wizards Everywhere? Film Tourism and the Imagining of National Identity in New Zealand." In *Tourism and National Identities: An International Perspective*, edited by Elspeth Frew and Leanne White, 52–63. Abingdon, England: Routledge, 2011.

Foote, Kenneth E. *Shadowed Ground: America's Landscapes of Violence and Tragedy*. Rev. ed. Austin: University of Texas Press, 2003.

Giraldo Velásquez, Claudia, Anne Marie Van Broeck, and Luisa Fernanda Posada. "El pasado polémico de los años ochenta como atractivo turístico en Medellín, Colombia." *Turismo y Sociedad* 15 (November 22, 2014): 101–14. https://doi.org/10.18601/01207555.n15.06.

Gunn, Clare A. *Vacationscape: Developing Tourist Areas*. 1972. 3rd ed. London: Taylor & Francis, 1997.

Knudsen, Britta Timm, and Anne Marit Waade. *Re-Investing Authenticity: Tourism, Place and Emotions*. Bristol, England: Channel View Publications, 2010.

Lai, Kun, and Xiang (Robert) Li. "Tourism Destination Image: Conceptual Problems and Definitional Solutions." *Journal of Travel Research* 55, no. 8 (November 1, 2016): 1065–80. https://doi.org/10.1177/0047287515619693.

Lennon, John J. "Dark Tourism Visualization: Some Reflections on the Role of Photography." In *The Palgrave Handbook of Dark Tourism Studies*, edited by Philip R. Stone, Rudi Hartmann, A. V. Seaton, Richard Sharpley, and Leanne White, 585–602. London: Palgrave Macmillan, 2018.

Lozanki, Kristin. "Desire for Danger, Aversion to Harm: Violence in Travel to 'Other' Places." In *Tourism and Violence: New Directions in Tourism Analysis*, edited by Hazel Andrews, 33–47. London: Routledge, 2016.

Munar, Ana María, and Szilvia Gyimóthy. "Critical Digital Tourism Studies." *Tourism Social Media: Transformations in Identity, Community and Culture*, edited by Ana María Munar, Szilvia Gyimóthy, and Liping Cai, 245–62. Bingley, England: Emerald Group, 2013.

Rincón, Omar. "Narco.estética y narco.cultura en Narco.lombia." *Nueva Sociedad*, no. 222, July–August 2009. http://nuso.org/articulo/narcoestetica-y-narcocultura-en-narcolombia/.

Roesch, Stefan. *The Experiences of Film Location Tourists*. Bristol, England: Channel View Publications, 2009.

Seaton, A. V. "Encountering Engineered and Orchestrated Remembrance: A Situational Model of Dark Tourism and Its History." In *The Palgrave Handbook of Dark Tourism Studies*, edited by Philip R. Stone, Rudi Hartmann, A. V. Seaton, Richard Sharpley, and Leanne White, 9–31. London: Palgrave Macmillan, 2018.

———. "Guided by the Dark: From Thanatopsis to Thanatourism." *International Journal of Heritage Studies* 2, no. 4 (1996): 234–44.

UNODC (United Nations Office on Drugs and Crime). "Estudio exploratorio descriptivo de la dinámica delictiva del tráfico de estupefacientes, la trata de personas y la explotación sexual comercial asociada a viajes y turismo en el municipio de Medellín, Colombia." Alcaldía de Medellín y Oficina de las Naciones Unidas contra la Droga y el Delito, Medellín, 2013.

Van Broeck, Anne Marie. "'Pablo Escobar Tourism'—Unwanted Tourism: Attitudes of Tourism Stakeholders in Medellín, Colombia." In *The Palgrave Handbook of Dark Tourism Studies*, edited by Philip R. Stone, Rudi Hartmann, A. V. Seaton, Richard Sharpley, and Leanne White, 291–318. London: Palgrave Macmillan, 2018.

Zavala, Oswaldo. *Los cárteles no existen: Narcotráfico y cultura en México*. México DF: Malpaso, 2018.

CONTRIBUTORS

FERNANDO ARMAS ASÍN is professor of history at the University of the Pacific in Lima, Peru. He is the author of *Una historia del turismo en el Perú: El estado, los visitantes y los empresarios, 1800–2000* (Universidad de San Martín de Porres, 2018). His scholarly interests include the history of religion, cultural heritage, and tourism in Peru.

RODRIGO BOOTH is a historian dedicated to the study the history of architecture, cities, and landscapes. He is an associate professor in the Architecture Department of the University of Chile. He has been a visiting professor and researcher at the National University of Rosario (Argentina), the National Autonomous University of Mexico, the Université de Strasbourg, the Université de la Sorbonne Nouvelle Paris 3, the Fondation Maison des Sciences de l'Homme, and the École des Hautes Études en Sciences Sociales in Paris. Booth has led research projects on cultural aspects of architecture, cities, and landscapes and has also published numerous articles on the history of mobility, history of tourism, history of landscape and infrastructures, and urban history and architecture in scientific journals from various countries. He authored the book *Luis Ladrón de Guevara: Fotografía e industria en Chile* (Pehuén, 2012) and is working on a book dedicated to the history of aesthetic notions of landscape in Patagonia between the mid-nineteenth and mid-twentieth centuries. He is director of the Art and Architecture Studies Group at the National Research Council of Chile.

FÉLIX MANUEL BURGOS is a social linguist with specialization in Latin American studies. His interests include the representation and construction of different social groups in mass media discourse, and he is currently studying varieties of the Spanish language used by news anchors working for *Noticiero Univisión* in the United States. Burgos has examined news reporting in his native Colombia during the armed conflict of 1998–2002.

MERI L. CLARK is professor of history at Western New England University in Springfield, Massachusetts, where she teaches Latin American and world history. She travels with her family to explore the world and cross-country ski. She has published several articles and chapters on nineteenth-century Colombian education and Latin American positivism.

ROCIO GOMEZ is assistant professor of Latin American history at Virginia Commonwealth University. Her research interests include Latin America, the history of science, environmental history, and the history of medicine. Her first book, *Silver Veins, Dusty Lungs: Mining, Water, and Public Health in Zacatecas, 1835–1946*, is published by the University of Nebraska Press. She is the 2019 RMCLAS recipient of the Edwin Lieuwen Award for the Promotion of Excellence in the Teaching of Latin American Studies.

KENNETH R. KINCAID is associate professor of history at Purdue University Northwest, where he has worked since 2007. He earned his PhD in Latin American history at the University of Kansas in 2005. His regional focus is the Andean republics, and he has published on invasive species, sacred space, indigenous traditions, immigration, and Lake San Pablo in northern Ecuador. He teaches classes on Latin American and global history.

ELIZABETH MANLEY is associate professor of history at Xavier University of Louisiana. She is the author of *The Paradox of Paternalism: Women and Authoritarian Politics in the Dominican Republic*

(University Press of Florida, 2017) and coauthor of *Cien años de feminismos dominicanos* (AGN, 2016) with Ginetta Candelario and April Mayes. She has published articles in *The Americas*, the *Journal of Women's History*, and *Small Axe*, is a contributing editor for the Library of Congress's *Handbook of Latin American Studies*, and is the co-chair of the Haiti–Dominican Republic section of the Latin American Studies Association.

MARK RICE is a historian of modern Latin America with a focus on Peru and the history of tourism. He is currently an assistant professor of history at Baruch College of the City University of New York. He is the author of *Making Machu Picchu: The Politics of Tourism in Twentieth-Century Peru* (University of North Carolina Press, 2018). His research on tourism and Latin American history has also appeared in edited anthologies and academic journals, including the *Radical History Review* and *Journal of Latin American Studies*.

ANADELIA ROMO received her training in history at Harvard University and is an associate professor of history at Texas State University. Her first book, *Brazil's Living Museum: Race, Reform, and Tradition in Bahia* (University of North Carolina Press, 2010), examines the discourse among Bahian intellectuals and state officials from the abolition of slavery in 1888 to the beginning of Brazil's military regime in 1964 and uncovers how the state's nonwhite majority moved from being a source of embarrassment to being a critical component of Bahia's identity. Her current project addresses the boom in illustrated tourist guides for Salvador from 1940 to 1970 and the way that the genre envisioned the city, its inhabitants, and questions of race and identity.

BLAKE C. SCOTT is an assistant professor of international studies at the College of Charleston in South Carolina. He is currently writing a book that examines the historical intersection of imperialism, identity, and tourism in the Caribbean. He has also published about tourism's history and culture in journals such as *Environmental History*, the *Journal of Tourism History*, and the *Caribbean Writer*.

EVAN WARD is associate professor of history at Brigham Young University. He received his PhD from the University of Georgia in

2000 and is the author of two books: *Border Oasis: Water and the Political Ecology of the Colorado River Delta, 1940–1975* (University of Arizona Press, 2003) and *Packaged Vacations: Tourism Development in the Spanish Caribbean* (University Press of Florida, 2008).

ANDREW GRANT WOOD is a historian, musician, and DJ who teaches at the University of Tulsa. His coedited volume (with Dina Berger) on Mexican tourism is titled *Holiday in Mexico: Critical Reflections on Tourism and Tourism Encounters* (Duke University Press, 2010). He is the author of *Agustín Lara: A Cultural Biography* (Oxford University Press, 2014) and is writing a book on the colonial history of the port of Veracruz, Mexico.

INDEX

Page locators in italics refer to illustrations.

advertising, 3, 8, 76; in Chile, 127, 130, 136, 140; about Dominican Republic, 249, 251, 253, 257–62, 265, 268, 271n24, 272n40; Latin American music and, 107n7, 110n64; for Narco-Tours, 305; Pan American Union (PAU) and, 86–88, 90, 93, 96, 103; about Peru, 49, 53, 55, 57, 60, 62–63, 154–55, 157, 160–62, 164; about Veracruz, Mexico, 171, 175, 180, 190n7
Africa, 3, 5, 9, 27, 91, 94, 98, 100, 178
airlines, 1, 11, 75, 154, 225, 254, 257, 267, 270n15; Aviateca, 223, 235, 238; Pan American, 6, 11, 67, 68, 69, 71, 74, 75–78, 251
airplanes, 71, 73, 75–79, 223, 233–34, 238, 240, 247
airports, 172–73, 229, 233, 238, 240, 254
alcohol, 29, 132, 205, 206, 208
Alemán Valdés, Miguel, 178, 180, 183
Amazon, 197, 200
ambassadors. *See* diplomacy
Arbenz, Jacobo, 229–31, 241
archaeology, 12, 58, 157, 162, 223–29, 232, 234, 239–40, 258–59
architecture, 3, 53, 128, 155; colonial, 13, 55, 157, 174, 226, 255, 259, 264, 280; pre-Hispanic, 55, 223–24. *See also* archaeology

Arévalo, Juan José, 228–31, 241
Argentina, 7, 62–63, 88, 90, 99, 107n8, 120, 151, 161, 200
Asia, 3, 5, 9
autobus, 177, *181*
automobiles, 3, 7–8, 80, 131, 160; Peruvian Touring and Automobile Club, 150–51. *See also* highways
aviation. *See* airlines; airports; flying

Balaguer, Joaquín, 245–47, 249, 253, 259, 263–64, 266–69, 270n10, 271n25, 273n54, 274n66
banks, 103, 154, 199, 236, 240, 257–58, 263, 265–66, 271n24, 271n28, 271n32, 274n66. *See also* World Bank
beaches, 9, 12, 119–21, 123, 125, 127–28, 141, 186, 214; in Dominican Republic, 245, 251, 255, 259–60, 264, 266, 274n66; in Veracruz, Mexico, 171–73, 175, 178
beach resorts, 1, 7, 127–28; in Dominican Republic, 245, 249, 253; Izabel, Guatemala, 239–40; social status of, 136–42; Villa del Mar, Mexico, 174, 176, 180, 182–83, *184*, 186, 193n51; Viña del Mar, Chile, 11, 128–29, 130, *131*, 132–35

beauty pageants, 260–64, 267–68, 272n40, 272n42, 272n46
Belize (British Honduras), 76, 228
boats and boating, 29–32, 51, 53, 64n12, 69, 77, 175, 178, 184, 257, 264. *See also* cruises; steamships
Bogotá, Colombia, 25–27, 197, 303
Bolivia, 9, 51–52, 55
Brazil, 9–10, 90, 94–95, 98–99, 103, 106n6, 120, 175, 200, 300
business: and "businesspeople," 5, 51, 61, 65, 79; in Chile, 132, 135, 141; in Dominican Republic, 251, 257, 262, 271n32; in Guatemala, 227, 231; interests, 102; knowledge of, 15; local impact of, 13; in Medellín, Colombia, 307; tour companies, 3, 6, 51–52, 60–62, 64n23, 101, 225, 227, 245, 305, 307; in Veracruz, Mexico, 171, 173, 175–77, 183–84; in Zacatecas, Mexico, 283, 294–95. *See also* development; markets

cafés, 42n25, 182–83. *See also* restaurants
California, 3, 61
camping, 153, 238
Canada, 5, 25, 172, 189, 200, 286
Cárdenas, Lázaro, 96, 102
Caribbean: beauty pageants in, 260–62, 272n42; and Colombia, 27–28; Dominican Republic's location in, 252; early aviation and, 71, 72, 73, 74; financial investment and development in, 245, 248; flights from, 172; hospitality culture in, 189; hotel and tourism industry in, 7, 263–68; increased travel to, 5–6, 9, 67–69, 79, 172, 224, 236, 239, 247–48; and international cooperation, 63; Pan American Airways and, 75–78, 251; plantation tourism in, 9; postwar growth of, 8; promoting images of, as paradise, 249; tourism literature about, 246, 248, 250–51, 273n63; travel conditions in, 70–71, 80–81; travel histories of, 10. *See also* Cuba; Dominican Republic; Puerto Rico
carnival. *See* festivals
Castillo Armas, Carlos, 231, 233–34, 241
Catholicism, 43n42, 99, 204. *See also* churches; church officials and clergy

celebrities, 50, 171, 184
Central America, 6, 89, 94, 172, 260–61, 265, 272n42. *See also* Guatemala
Chile: El Recreo beach, 124; industry in, 90; international cooperation and, 63; international diplomacy and, 200; landscapes in, 139; Ministry of Education in, 96, 110n51; Pan American Union delegate from, 85; Santiago, 6, 56, 140; in South American tours, 51, 62; state investment in tourism in, 151–52, 163; steamship companies and, 51; Subercaseaux Mackenna family from, 128–29; tourism and health in, 123; tourism literature and, 8; trade, 85; travel and leisure among elites in, 125–29; travel guides about, 140; Valparaíso, 51, 129; writings on indigenous population in, 110n55. *See also* Viña del Mar, Chile
churches, 53, 55, 60, 174, 255, 266
church officials and clergy, 56, 204, 223, 265
climate, 31, 37, 58, 61, 127, 155, 171, 175, 187, 260
coffee, 16, 260
Colombia: and Alberto Lleras Camargo, 102; Cartagena, 303; and Colombian-German Air Transport Association, 75; and dark tourism, 9, 14, 300, 303, 304, 305, 306–9; and drug cartels, 299, 304, 307, 308, 311, 312, 313; industry potential in, 91; journal distribution in, 41n7; landscape of, 23; "Macondo," 5; and *Narcos* series, 300, 302, 304, 308, 309, 310, 311, 312–13; national identity in, 34, 39–40; nineteenth-century writers from, scholarship on, 41n11; peace agreement in (2016), 303, 313; prostitution in, 304; and Robert Escobar, 307; Samper and Acosta in, 25–31; trade with Ecuador, 202. *See also* Bogotá, Colombia; Escobar, Pablo; Medellín, Colombia
Columbus, Christopher, 246, 248–49, 252, 254–55, 257–58, 264–66, 269, 274n71, 274n72
communication infrastructure, 1, 24, 49, 90. *See also* radio; telephone; television
communism, 201, 230–31

conferences: Commercial Aviation, 100; First American International, 85; Foreign Ministers, 200; Inter-American (1959), 195–97, 199–203, 211–12, 214, 216; Inter-American Indian, 97, 110n53; Inter-American Travel, 87–88, 92–93, 99–101, 103, 106n3, 106n5, 109n35, 111n66, 111n70; Pan American, 76; Pan American Child, 107n7; Pan American Highway, 100; Pan American Scientific, 56; Pan American Union, 93, 110n53, 111n70; Peru national tourism, 162
congress: Colombian, 26–27; Ecuadorian, 200; Peruvian, 161; U.S., 75, 294
Costa Rica, 76, 90
cruises, 1; Caribbean, 5–6; Central and South American, 61–62; Grace Line, 6, 51, 52, 60–62, 101–2, 225; Panama Canal, 6; Peru, 78; and shipping industry, 102, 225; South Atlantic, 51. *See also* steamships
Cuba: and Afro-Cubanismo, 98, 100; in *Bulletin of the Pan American Union (BPAU)*, 104, 108n17; and Casino and Tourism Bill, 7; and dark tourism, 9; government promotion of tourism in, 151, 161; hotels in, 7; international tourism to, 247; Pan American Union promotions for, 89; Pan Am flights to, 67, 69–70, 75–76; as protectorate of U.S., 73; Steven Crane travel to, 69–70; tension with U.S., 216; Tourism Commission, 7; and tourism in Guatemala, 235; travel to Mexico from, 172; Ward Line, 5
Cusco, Peru, 8, 14, 49, 50, 52, 55–56, 59–62, 95, 99, 152, 157, 158, 163

democracy, 28, 133, 161, 186, 212, 216, 241
development: agencies in Guatemala, 224; air transportation and, 68, 73, 74, 75, 77–78; Alliance for Progress and, 235, 238, 270n23; challenges to, 61–62, 129, 141–42, 178–79; coordination of, 62; Cuban 1919 Casino and Tourist Bill and, 7; Dominican Republic coastal resorts and, 245, 253, 259, 262–63, 266–69; Dominican Republic National Development Commission and, 251; Dominican Republic National Tourism Bureau and, 247, 250, 254, 260; Guatemalan archaeological sites and, 227, 229–30, 236; Guatemalan Petén region and Tikal site and, 230, 236–41; hotels and, 7, 8, 32, 103; impact of motor vehicles on, 8; infrastructure and, 3, 6, 173, 175; Inter-American Development Bank and, 103; Inter-American tourism and, 93; Mexican Automobile Association and, 7; Mexican Ministry of the Interior and, 173; modernization and, 103; Peruvian Ministry of Development and Public Works and, 152, 162; Peruvian National Tourism Corporation (CNT) and, 161–63; Peruvian tourism and, 147, 149, 151, 154, 161–65; promotion of national progress and, 25, 27, 89, 104, 250; regional economies and, 50, 59; resistance to, from native communities in Ecuador, 210, 212–13; steamships and, 71; United States Agency for International Development (USAID) and, 236; World Bank and, 271n23, 271n24. *See also* business
diplomacy, 25, 60, 102–3, 105, 174–75, 199, 234, 274n66
Dominican Republic: and *Bohío Dominicano*, 251–52, 258, 260–62, 264–68, 270n17, 272n46, 273n63, 273n65, 274n76; colonial heritage of, 246, 249, 250–51, 254–55, 258–59, 262–64, 266–67, 269, 271n28, 274n66; foreign consultants in, 246, 255, 262, 271n32; international funding for tourism in, 253; National Tourism Bureau, 245–46, 250–54, 260, 262–63, 268, 269n2, 272n40; tourism legislation in, 253. *See also* Balaguer, Joaquín; beauty pageants; Caribbean; Columbus, Christopher; fantasy; heritage tourism; Santo Domingo, Dominican Republic; Trujillo, Rafael; United Nations Educational, Scientific, and Cultural Organization (UNESCO); World Bank
drugs, 299, 300, 303–4

Ecuador, 10, 12, 30, 41n7, 51, 155; and Eleventh Inter-American Conference, 195–97, 199–203, 211–12, 214, 216; Lake San Pablo, 195, 196, 202, 203, 205, 210–13, 214, 215–16. See also indigenous peoples and cultures
employment, 1–2, 14, 78, 79, 153, 187, 188, 205–6, 257, 304
England: English Pacific Steam Navigation Company, 51; investment in Peruvian tourism, 10, 49, 52; Pan Am service to, 77; Samper and Acosta's impressions of, 24, 28–29, 34–38, 40; tourists from, in Chile, 137; travelers to Alpine destinations, 33, 36–38, 42n26, 42n34
environment, 2, 69, 73, 80, 121, 128–29, 132, 134, 138, 153, 279, 286, 292
Escobar, Pablo, 299–300, 301, 302–4, 305, 306, 308, 309, 310, 311–13
Europe: and aftermath of WWII, 78, 102, 229; architectural style in, 53; beach culture in, 120; challenges traveling in, 30; Chileans traveling in, 128–29; early air travel in, 75; film about Cusco screened in, 61; increased travel from, to Latin America, 89, 91–92; Pan Am Airways service to, 77; Samper and Acosta's travel to, 26, 29–35, 37–40; Swiss alpine tourism and, 31–34, 36, 38, 40; U.S. travelers to, 40–41n2. See also England; France; Grand Tour, European; Spain

fantasy, 247–50, 252, 254, 260, 262–63, 269
festivals, 26, 94–95, 99, 177, 179, 182, 189n1, 190n7, 192n43, 261, 274n71
fiction, 5, 246, 300, 302, 305, 308, 312. See also fantasy; film; literature
film, 61, 100, 187, 234, 300, 302, 305, 310, 312
fishing, 70, 155, 172, 174, 177, 189n3, 202, 226, 272n38
Florida, 7, 67, 69–70, 75, 75–76, 79–80
flying, 6, 67–69, 73, 76–80, 223. See also airlines; airports
"folk" culture: Dominican Republic exhibitions of, 254, 272n40; and folklore movement, 107n8, 109n40; Inter-American Travel Congress and, 100–101; interest in, in Latin America, 86, 94, 103–5; Mexican mining stories as, 288; in national tourism promotions, 98–99, 149; and Pan American Union, 88, 91–92, 94–99, 101, 110n55, 111n74; Peruvian consideration of, 152, 157, 163; and regional culture in Veracruz, Mexico, 98, 188; and Spanish popular culture, 28
France, 2–3, 24–25, 29–32, 32, 35, 128

Germany, 2, 24, 33–34, 37–38, 75, 111n74, 137, 225
Global North, 246, 250, 258
Good Neighbor era, 86, 92, 108n33
Grace Line, 6, 51, 52, 60–62, 101, 225
Grand Tour, European, 2–3, 24, 33–34, 39, 89, 128
Great Depression, 7–8, 61, 86, 149–51, 165, 174
Guatemala: Counterrevolution of 1954 in, 224, 230–31, 241; early tourism in, 224–26; foreign consultants and financing in, 224, 228, 230, 236, 239; international agencies in, 224; International Exposition, 225; Izabal, 239–40; Petén region of, 224, 227, 228, 229, 230, 232, 233, 236, 238–39, 241; and Sylvanus Morley, 227–28; Tikal, 223–25, 227–31, 232, 233–34, 235, 236–37, 239–41; Tourism Commission, 231; United Fruit Company in, 224–25, 226, 228, 230. See also Arbenz, Jacobo; Arévalo, Juan José; Castillo Armas, Carlos; indigenous peoples and cultures; Ubico, Jorge
Guatemala City, 225–26, 231, 233, 235–37, 240
guides: to Chile, 135, 140; Chilean Vacationer's Guide, 8; Cook Agency, 3; early twentieth-century, of South America, 50, 58–59; Mexican Automobile Association, 7; Pan American Union publications and, 87–91, 98–99, 104, 107n11, 108n17, 112n84

Havana, Cuba, 5–7, 67, 71, 73, 74, 76, 79, 172
health, 3, 14, 136, 150, 152–53, 155, 175, 178, 180, 184, 247, 286

heritage tourism: and Dominican Republic, 248–50, 255, 258–59, 262, 267, 271n28; and Guatemala, 224–25, 227, 236, 241; and Mexico, 172, 174, 186, 280, 283, 296n2; Pan American Union acknowledgement of, 101. *See also* "folk" culture; indigenous peoples and cultures
highways, 6–8, 100, 151, 154, 236. *See also* automobiles; roads
horses, 30–31, 38, 55–56, 59, 126, 129, 135, 140, 172
hospitals, 174, 258, 274n66, 293
hotels: in Dominican Republic, 245, 251, 253–55, 257–58, 260–61, 265–66, 272n42, 273n53, 274n66; in Ecuador, 195–96, 199, 201–4, 209, 211–14, 216; in Guatemala, 230, 237–38; Hotel Company of Peru, 147, 148, 161–62; Hotel Society of Peru, 153; and Mexican Hotel Association, 175; in Peru, 53, 56–57, 60–62, 152–55, 156, 157, 158, 159, 160–65; Peruvian indigenous culture and, 160; in Switzerland, 32, 34, 36, 38, 42n34, 42n38; in Veracruz, Mexico, 172–73, 176, 177–80, 182–83, 184, 191n31; in Viña del Mar, Chile, 136–38, 141, 143n39

imperialism, 11, 71, 73, 81n116, 86, 105, 106n5, 228, 241
Inca civilization, 53, 55, 57, 61, 90, 95, 157
indigenous peoples and cultures: and Ecuadorian conflict over hotel-casino, 195–96, 202–10, 212–13, 215–16; and Ecuadorian indigenous land rights, 212–13, 215–16; festivals and ceremonies of, 95–96, 101, 160; and Guatemalan national identity, 229; and Guatemalan tourism, 225–26, 234, 241; and images of miners in Zacatecas, 285; and *mestizaje*, 27; national culture and, 86, 88, 94, 96, 98–99, 103, 149, 157, 229; Pan American Union and, 96–98, 102, 110n55; in Peru, 149, 150, 157, 160, 165; recognition of, 91–92, 94–96; specialists in the study of, 60, 95; tourism promotion and, 101, 102–3, 157, 160, 165, 286; U.S. government warnings about, 109n39. *See also* "folk" culture

industrial process and culture: and *Bulletin of the Pan American Union*, 88, 90; challenges of, and social reform, 147, 149–50, 165; goods from, 119; and growth of technology, 23–25, 34, 49; in Valparáiso, Chile, 121, 131–32, 134, 141; in Zacatecas, Mexico, 279–81, 286–87, 289–90, 295, 296n2
islands. *See* Caribbean; Cuba; Dominican Republic; Puerto Rico

jungle, 6, 80, 157, 229, 233–34, 237, 239

labor: conditions, 109n39; as contrast with leisure, 135; dark tourism and, 283, 287; denigration of, 126; mining, 288, 291–93; native, 202, 225, 241; reform in Peru, 165; time, 119; unionists in Mexico, 176. *See also* employment; workers; working class
lakes: San Pablo, Ecuador, 195–96, 202–6, 209–16, 218n26; in Switzerland, 31–32, 34, 40; Titicaca, Bolivia, 52–53, 55, 64n12, 157
legislation: Dominican Republic support for tourism, 252–53; for National Tourism Corporation in Peru, 161; need for equitable, in Ecuador, 196
Lima, Peru, 27, 31, 50–51, 53, 57–63, 110n53, 152–55, 157, 160–61, 163, 165
literature, 3, 10, 24, 26, 30, 33, 41n2, 49, 89, 98, 109n40, 140, 246, 305
London, England, 3, 23–24, 26, 56, 75, 199

Machu Picchu, 8, 50, 56–57, 57, 59, 147, 153, 239
magazines and journals: *Américas*, 103; *Art and Life*, 96; *Bohío Dominicano*, 251–52, 258, 260–62, 264–68, 270n17, 272n46, 273n63, 273n65, 274n76; *Bulletin of the Pan American Union*, 96, 97, 103, 107n12, 108n13, 111n74; *Caribbean Beachcomber*, 251, 270n15; *Cultura Peruana*, 148, 156, 157, 158, 159; in Ecuador, 207; *El Mosiaco*, 26; fashion publications, 121; *Fortune*, 73; *La Revista Nacional de Turismo*, 173; *Life*, 234–35; *Mexican Art and Life*, 94; in Mexico, 171, 173; *New Horizons*, 72; *Perú*, 60; *Peru To-Day*, 57–58; *Sucesos*, 130–31; *Turismo*, 151–53; vicarious travel through, 140; *Viña del Mar*, 143n44; *Zig-Zag*, 122, 134, 139. *See also* guides

mail, 5–6, 73, 75–76
markets: Alpine tourism, 24; Bolivian, 52; capitalism and, 8n3; Central American Common Market, 236; changing conditions of, 61; Colombian reputation in global, 313; European tourist, 32; mass tourism, 225; North American travel, 226; Peruvian national, 50, 63, 152, 160; silver, 294; travel, 3, 59, 63, 64n23, 236, 248. *See also* business; development; Great Depression
markets, local and regional: in Guatemala, 226; in Peru, 160
Medellín, Colombia, 300, 303–4, 306, 307, 308–9, 310, 311, 312, 313
medicine, 3, 52, 119, 272n46, 291, 293
Mexico: Acapulco, 171, 177, 180, 184, 186, 188; archaeological sites in, 227–28, 239; *Bulletin of the Pan American Union* coverage of, 90, 94–95, 99; cultural relations with U.S., 111n68; Department of Tourism in, 177; and Dominican "Friendship Weeks," 254, 271n25; early tourism in, 151, 172–73; economic growth in, during and after WWII, 175; Federal Improvement Association in, 178, 184; Guatemalan independence from, 241; indigenous and folk culture in, and tourism promotion, 98–99, 102; and Inter-American Indian Institute, 110n53; Inter-American Travel Congress hosted in, 101; key tourist destinations in, 177; postwar tourism boom in, 172; reimagining national identity in, 94, 98; U.S. auto caravan to, 111n70. *See also* archaeology; Mexico City; Veracruz, Mexico; Zacatecas, Mexico
Mexico City, 2, 6, 111n70, 171–72, 175, 177, 179, 181, 181–83
Miami FL, 2, 6, 76–77, 79, 82n24
middle class, 3, 24, 33–35, 42n26, 59, 63, 79, 126, 171, 263
military, 9, 71, 73, 75, 78, 92, 106n5, 150, 164, 200, 240, 294, 309, 319
mining, 200, 280–81, 283–95
mountains. *See* Europe
mules, 5, 31, 37–38
music, 3, 73, 92, 94, 97–98, 105, 174, 313

national identity: dark tourism in Colombia and, 300, 312–14; folk music and dance and, 96, 98, 107n8, 110n64; postwar consolidation of, 103; Samper and Acosta on, 23, 28–29, 34; São Paulo 1922 Modern Art Week and, 94; tourism in Peru and, 147, 149–50, 154–55, 157, 165. *See also* "folk" culture; indigenous peoples and cultures
national parks, 2–3, 233, 235, 237–41
navigation, 51, 69, 70, 77
New York City, 2, 5–6, 51, 56, 60–61, 79, 102, 143n44, 231, 265, 273n65, 299
Nicaragua, 8, 76, 223
nightclubs, 104, 171–72, 180, 183, 266, 276n66, 285
North America, 49–51, 53, 62, 64n23, 138, 226, 247, 281. *See also* Canada; Mexico; United States

Organization of American States (OAS), 87, 102–4, 106n1, 111n76, 196, 197, 201, 216, 224, 239, 253

Panama, 6, 26, 51, 53, 62, 76, 79, 94
Pan American Highway, 6–7, 100, 154, 236
pan-Americanism, 56, 76, 90, 92, 97, 100, 106n6, 107n7, 110n64
Pan American Union (PAU), 11, 61, 85, 87–88, 96, 97, 103, 105, 105n1, 107nn10–11, 109n49, 111n76. *See also* magazines and journals
Pan American World Airways, 6, 11, 67–69, 71, 74, 75, 223, 251
Paris, France, 2, 23–26, 29–31, 34, 37, 75, 128, 155
Peru: Arequipa, 50–53, 55–56, 61, 154–55, 157, 163; Great Depression and, 61–63, 151–52; Hiram Bingham and, 56, 57, 59; and Peruvian Corporation, 52–53, 58–59, 154; and Peruvian Touring and Automobile Club (TCAP), 60–62, 150–51, 153, 160; Populism era in, 149–50; and tension with Ecuador, 200. *See also* Cusco, Peru; "folk" culture; hotels; indigenous peoples and cultures; Machu Picchu

photography, 59, 128, 134, 139, 171, 261, 272n44, 285, 300, 305, 309
poetry, 22, 31, 39, 130, 227
police and policing, 35, 132–34, 141, 195, 204–6, 208, 246
Ponce Enríquez, Camilo, 197, 199–200, 207, 214, 218n5
ports, 5, 7, 12, 32, 50–52, 62, 77, 79, 90, 119, 121, 141, 171–89, 189n1, 225, 233, 247
Portugal, 30, 71, 98
postcards, 59–60, 138–39
Prado, Manuel, 150, 154, 164
public health, 14, 150, 175, 178, 180, 184. *See also* sanitation
Puerto Rico, 7, 9, 88, 247, 254, 270n15, 271n25

Quito, Ecuador, 41n7, 196–97, 199, 201–2

"race" and racial groups, 96, 98, 149–50, 195–96, 202; "whites" and, 100, 111n69, 160–61, 202, 204–5, 208, 211–12, 214, 216. *See also* "folk" culture; indigenous peoples and cultures
radio, 77, 162, 187, 204, 207
railroads: between Argentina and Chile, 62; Latin American dignitaries tour of U.S., 85; Mexican National Railways, 102, 176–77; from Mexico City to Veracruz, 172, 175, 177; Miramar station (Chile), 129, 131–32, 136; in Peru, 54, 55, 57–58, 60; and Peruvian Corporation, 52–53, 58, 58–59, 154; Peruvian Railway, 61; Peruvian Southern Railroad, 52; in Spain, 30; in Switzerland, 32, 34, 37; and transformation in travel, 5, 8, 24, 42n22, 80; United Fruit line in Guatemala, 225; in Valparaíso, Chile, 121, 135
restaurants, 127, 138, 180, 183
Rio de Janeiro, Brazil, 6, 91, 94, 200
rivers, 6, 27–28, 32, 39–40, 52, 213, 274n72, 285, 286
roads, 5, 8, 30, 32, 56, 154, 173–74, 214, 232, 236–37, 240. *See also* highways
ruins. *See* archaeology; mining

sanitation, 7, 12, 141, 175, 178–80, 191n31
Santo Domingo, Dominican Republic, 251–52, 254–55, 257–59, 261, 266–67, 271n28, 274n66
social reform, 149–53, 161, 164–65, 230–31, 308
social welfare, 151, 153, 197, 204, 211–15
South America, 4; Charles Darwin and, 71; coastal recreation and development in, 120; dances of, 96; early foreign travel to, 49–51; first commercial airline in, 75; Grace Line and, 51, 102; growing traveler focus in, 89; international travel companies in, 61–62; miners in, 291; "most beautiful train ride" in, 55; Samper and Acosta and, 27, 30, 35, 37, 39–40, 41n7; *Seeing South America*, 91. *See also* Argentina; Bolivia; Brazil; Chile; Colombia; Ecuador; Peru
souvenirs, 35, 59, 85
Spain: colonial power of, 71; Columbus and European conquest, 248–49, 255, 257, 269; Crown and silver, 279, 290; dance and, 96; *encomenderos* from, in Ecuador, 212; and first colonial settlement in Caribbean, 246; George Santayana and, 309; Grand Tour and, 24; Guatemala's independence from, 240; and ownership of silver mines in Mexico, 294; Samper's travels in, 28–30
steamships: arriving in Dominican Republic, 247; compared to air travel, 73; early companies for, 5; *Good Neighbor Tour* and, 92; after Great Depression, 61; impact of, on travel, 24, 49, 71; increasing presence of, 30; on Lake Titicaca, 55; on Magdalena River, 27; Panama Canal and Pacific coast service, 89–90; service to Peru, 51–52, 52; tonnage of, 88; and tourism in Veracruz, Mexico, 172; and travel in Switzerland, 31–32, 34; United Fruit vessels, 224. *See also* boats and boating; cruises; Grace Line
swimming, 119, 121, 125, 129, 137, 140, 141, 174, 178, 184, 186
Switzerland, 2, 3, 23–25, 30–32, 36, 40, 42n24, 226

technology, 5–6, 11, 34, 49, 71, 73, 77–80, 287, 291–95. *See also* communication infrastructure

telephone, 90, 252, 309
television, 268, 283, 299–306, 308–14
tourism industry: Alliance for Progress and, 224; contemporary realities of, 1–2; dark tourism and, 283; Great Depression and, 7; growth before WWII, 6–8, 49–51, 61–63, 98–99, 101, 149, 151, 247–48; and increased opportunity for Inter-American exchange, 93–94; Latin American economies and, 151; and recreational beach development, 120; and rise of air travel, 75–79; social and ethical concerns of, 14–16; and U.S. spending on foreign travel, 91; and visitor spending in Colombia, 303. *See also* business; development
tours: Argentina to Chile rail, 62; favela, 9; Guatemala, 225, 234, 241; Medellín, Colombia, 299–300, 303–14; Mina El Edén, 279, 281, 283–86, 288–93; San Juan de Ulúa fortress, 186; Thomas Cook, in Lima, 62. *See also* cruises; Grace Line; Grand Tour, European; guides; travel agencies
trade: air travel and, 73; fair, 16; Pan American Union and, 85, 88–90; transoceanic, 77
transportation, 3, 24, 34, 67–68, 71, 75, 100, 111n, 172, 174, 225, 257. *See also* airplanes; automobiles; steamships
travel agencies: Thomas Cook and Sons, 3, 61; Wagons-Lits, 61–62; Wagons-Lits Cook, 62, 64n23
Trujillo, Rafael, 245, 247, 269n1, 272n40, 273n63

Ubico, Jorge, 225, 228–29, 230
United Fruit Company, 5, 7, 224–25, 226, 228, 230; Great White Fleet, 5, 225–26
United Nations Educational, Scientific, and Cultural Organization (UNESCO), 70, 224, 253, 255, 271n28
United States (U.S.): Alliance for Progress, 103, 224, 235–38, 241, 252, 270n19; aviation industry, 75–76, 78–80, 81n17; CIA *coup d'etat* in Guatemala, 230–31; screening of *Cuzco, el Imperio de los Incas* in, 61; Drug Enforcement Agency (DEA), 303, 312; early tourism in, 3; leisure class in, 126; occupation of Dominican Republic, 245, 271n25; Pan American Union, 85–87, 89–90, 92–93, 95, 98–103, 105, 106n5, 107n10, 110n57, 111n70; passenger flights, 67, 69, 73, 172; postwar tourism industry growth, 189, 224; Rio de Janeiro Protocol, 200; Silver Purchase Act, 294; tensions with Cuba, 216, 235; travelers, 40n2, 49, 51, 228, 247–48, 250
United States Agency for International Development (USAID), 224, 236–37, 239–40, 241
United States Department of Commerce, 90, 93

Velasco Ibarra, José María, 197, 200
Venezuela, 30, 41n7, 85, 196
Veracruz, Mexico: carnival celebration, 177, 179, 182, 190n7; Club de Regatas, 184; Mexican Hotel Association meeting in, 175–77; sanitation campaigns, 175, 180; San Juan de Ulúa, 173, 180, 183, 186; Semana Santa, 172, 179, 186–88; Toña la Negra, 184; tourism marketing and promotion of, 173, 175, 177, 182–83, 190n7; urban infrastructure, 173–75, 179, 184, 189; U.S. vice consul description of, 174–75; Veracruz Hotel Association, 179. *See also* Alemán Valdés, Miguel; beach resorts; fishing
Viña del Mar, Chile, 7, 11, 120, 129, 130, 139, 140; dangers in, 132; elite social gatherings in, 134–37, 140; founding of, 131; Gran Hotel, 136–37; Hotel Miramar, 138; Miramar beach, 120, 123, 132; municipal government in, 132–33; postcards of, 139
volcanos, 53, 202, 210–11

Washington DC, 61, 76, 86, 227–28
workers, 14, 75, 131, 149–50, 153, 161, 208, 212, 233, 279–80, 283–95. *See also* employment; labor

working class, 59, 126–27, 132, 134, 163, 165
World Bank, 224, 240, 253, 271n24
World War I, 75, 92
World War II, 6, 8, 77–78, 86, 88, 92, 100, 165, 172, 175, 228

Zacatecas, Mexico: industrial heritage of, 279–81, 286–87, 289–90, 295; Mina El Edén, 279, 281, 283, 285–86, 288–89, 291–93, 295; and silver markets, 294; silver mining in, 281, 284, 290–92, 295

www.ingramcontent.com/pod-product-compliance
Lightning Source LLC
Chambersburg PA
CBHW031900220426
43663CB00006B/698